D0991174

Home Security

Home Security

Alarms, sensors and systems

Second edition

Vivian Capel

Newnes
An imprint of Butterworth-Heinemann
Linacre House, Jordan Hill, Oxford OX2 8DP
A division of Reed Educational and Professional Publishing Ltd

A member of the Reed Elsevier plc group

OXFORD BOSTON JOHANNESBURG
MELBOURNE NEW DELHI SINGAPORE

First published 1994
Reprinted 1994
Second edition 1997

British Library Cataloguing in Publication Data
A catalogue record for this book is available
from the British Library

ISBN 0 7506 3546 0

Library of Congress Cataloguing in Publication Data
A catalogue record for this book is available
from the Library of Congress

Printed and bound in Great Britain by
Biddles Ltd, Guildford and King's Lynn

Contents

Preface

As the family wagon rolled to a halt you felt that on the whole it had been quite a good day. Pity about the ham sandwiches though; really should have known better than to put the picnic hamper in the back with the dog. It had been a mistake to let offspring no. 2 have that last ice-cream; never mind, the back seat needed a good clean anyway. The nose-to-tail crawl back from the coast had been tiresome, but now, duty done, you are home at last, and ready for a nice relaxing feet-up-before-the-telly evening to savour the bank holiday TV offerings.

But on opening the front door, you see that things are clearly far from what they should be. Doors are open that should not be open. There is no welcoming blink from the video timer in the living room, because there is no video, just blank wall – as there is where the TV should have been, and the hi-fi, and offspring no. I's games computer.

A dash upstairs to the bedroom reveals a scene of utter devastation. Every drawer has been turned upside-down, spewing its contents into a great mound on the floor. Clothing and other items have been hauled out of cupboards to join the heap. The dressing table has been cleared of jewellery and trinkets. An obscene message has been scrawled on the mirrors with lipstick.

Numbness and incomprehension soon give way to anger and outrage as you realize that many irreplaceable treasures have gone: grandfather's retirement presentation gold watch; grandmother's ruby engagement ring; that inlaid ivory snuff-box that had been in the family for generations, which was thought to be priceless, and which one day you intended to get valued. All probably right now changing hands in some pub for a few paltry pounds.

By now many readers will be sadly nodding their heads as they recall their own similar unhappy experience. Yet you may have been fortunate; some have returned home to find their entire house cleared, including carpets!

Soon you will have discovered that insurance is no real compensation. It can never make up for the long heart-breaking hours spent cleaning up, sorting out, searching, agonizing over lost treasures, listing, evaluating, and trying to obtain replacements. Then there are those missing items that will be remembered in the weeks and months ahead when it is too late to claim.

Perhaps worst of all is the trauma experienced by the womenfolk, knowing that some yob's fingers have been pawing over their personal and intimate belongings. They feel violated almost to the point of rape. That sense of privacy and home security has been shattered. Every holiday and evening out for years to come will be overshadowed with the memory of this day, and the dread of what might be awaiting on returning home.

Some feel that the only way to exorcize the trauma is to move. But this is no answer: no area is free from crime. Some have moved and been burgled in their first week; new residents are often prime targets.

There is not even consolation in the hope that lightning does not strike twice in the same place, that once done you won't be burgled again. Often the villains come back after giving you time to get new replacements. They now know the layout of your house and can expect a newer, more valuable haul that the first.

If such an experience has not happened to you, you are very fortunate. With burglary increasing at an alarming rate, your turn will surely come, *unless you take positive steps to guard your home!* There is a double reason to do so: in addition to the increasing crime rate, more householders are improving home security and installing protective devices. So criminals are seeking out the easier jobs, the unprotected ones – one of which could be yours!

The problems facing most ordinary folk are as follows. What measures can and should be taken? Which are the most effective? What do all those confusing technical terms quoted with alarm systems really mean? Could I install one, or must it be done at no little expense by professionals? Is in fact an alarm the complete answer?

Large sums can be and often are spent unnecessarily on security, yet weak links are frequently left unrecognized – except by the burglar! Advice sought from shops that sell security items is often either not forthcoming, or downright misleading. But while good advice is usually available from specialist security firms, they are understandably prejudiced in favour of the items they stock, and obviously can't tell you everything you need to know.

Be assured, though, that anyone with moderate DIY skills can make his or her home safe and secure to an acceptable degree, and install an effective alarm system. *You* can do it with the necessary guidance.

The purpose of this book is to give that practical guidance. It shows the burglar's preferred methods of entry, the weak points he looks out for, many of the tricks he uses, and what you can do to thwart him. It also explains the pros and cons of alarm systems, discusses which features are important, and clears away the mystery surrounding those technical terms and expressions. It shows how to install a system, how to prevent faults, and what to do if any occur. Other security devices are given a critical scrutiny.

For readers having some technical and constructional skills, a description and circuits are given for two simple and tried designs for control units. One is for a domestic system and the other for a foolproof public-hall system. These have many advantages over most commercial units, and are well worth building for those able to do so.

Vivian Capel

Acknowledgements

Equipment described in the text can be obtained from the following:

- Camrex Coatings
- Celtel
- Cricklewood Components
- Direct Choice
- Innovations
- Maplin Electronics
- Tann Synchronome

Acknowledgements

1 Physical security

It is surprising the number of people who install expensive alarm systems and yet neglect basic physical security measures. A house having an alarm but offering easy entry may tempt a burglar to chance it, snatch a few valuables, then make a quick getaway. He may gamble on the fact that because of the prevalence of false alarms, few people take any notice of an alarm sounding unless it goes on for too long. He may even try, and succeed, in silencing it.

The alarm system should always be regarded as the second line of defence, and reliance should never be placed solely upon it, however good it may be. The first and main defence is a perimeter that is as solid and impenetrable as it can reasonably be made. Before an alarm system is even considered, the physical security of all possible access points must be assessed, and improved if found below standard.

This is becoming increasingly important with the incidence of 'smash-ins' in which intruders smash their way through doors regardless of noise, while the occupants are at home and likely paralysed with fear. They take a few valuables, usually the TV and video, and are gone, all within two or three minutes. An intruder detection system is obviously of no help, although an alarm system having a panic button could be. However, substantial physical protection could delay the thugs sufficiently to make a 999 telephone call, or deter them enough to make them go elsewhere where entrance is easier.

Even so, while a good level of security is desirable, it can be too high. Unnecessary security can not only be a waste of money but, what is even worse, impose irksome restrictions and limitations on everyday life. The occupants could feel that they are living in a fortress. Remember that the security system must be lived with month in and month out, year in and year out, for years to come. The goal should therefore be *friendly security* – friendly that is to the occupants.

Most break-ins are opportunist, often by youths or even children. If they find a weak point they will take advantage of it, but if not they are likely to try elsewhere. Why waste time and effort cracking a hard nut when there is one with a soft shell around the corner?

Of course, if your premises are known or believed to contain articles of high value, they may attract the attention of the professional burglar. He will not be so easily deterred, and a much higher level of security will be required; the added inconvenience must then be accepted.

In general, security arrangements should cause as little inconvenience as possible, although some is inevitable. The danger with irksome over-security is that sooner or later it is relaxed and neglected, thus making the premises more vulnerable than they would be with more moderate security that is kept up. The important thing is to identify and protect the vulnerable points while being less fussy over low-risk areas. Also if there is more than one way of achieving a similar degree of protection, the method should be chosen that will be the least inconvenient to operate and use. This will be a significant factor in our discussion of various security measures.

Modus operandi

We will first briefly take a look at the burglar's *modus operandi* – the things he looks for, how he gets in, and what he does when he's in.

The favourite method of access is a rear window. He is less likely to be observed at the back than at the front, or be caught by someone returning home at the front door. Furthermore, people are often careless over security at the back of the house. Frequently, front doors can be seen fairly bristling with high-security locks and bolts, while rear doors have just a simple lock, and windows have virtually no protection at all. It is not unusual for rear windows to be left open, which is as good as displaying a 'burglars welcome' sign.

Windows or doors that are screened from public view by high walls, bushes, or trees are especially welcomed by the burglar, as are darkened areas away from street lighting.

After breaking in, the burglar's first task is to establish an exit route by opening an exterior door. This enables him to carry out bulky items and also to make a quick escape if he should be disturbed. If he is unable to do this he is forced to leave the way he came, usually through a window. This may be difficult, so he feels trapped and vulnerable and is unlikely to stay long, or remove large objects. The blocking of all possible exit routes is thus an important line of defence which will minimize loss should an entry be made.

Usually the exit chosen is the back door, and often the burglar will bolt the front door, if bolts are fitted, to prevent anyone coming home and catching him. Securing the back door against both entry and exit is thus vital in thwarting the preferred exit route.

Bolts on the front door are therefore of dubious value: they cannot be used when the premises are vacated, yet they can be used by the intruder to keep you out. While they increase security at night, they reduce it at other times. One solution would be to withdraw and conceal removable bolts each morning and replace them at night.

Having taken that brief look at how the burglar operates, we will now

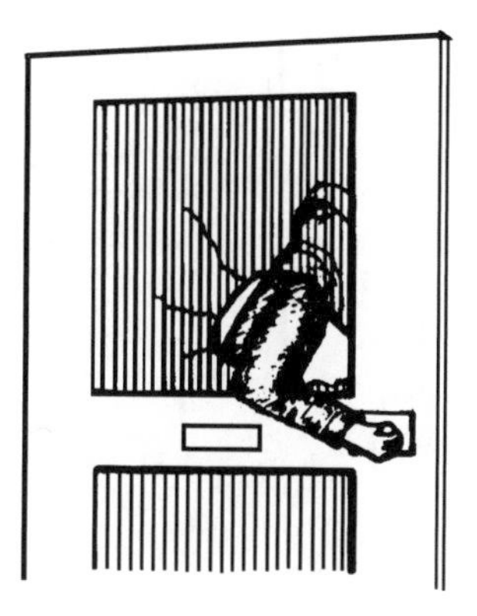

Figure 1 *Glass-panelled front doors are extremely vulnerable, especially if ordinary locks are fitted. Such doors should always be wired to the alarm. Solid wooden doors with security locks need not be wired as these are not likely to be attacked. All back doors into the main building should be wired.*

consider more closely the individual access points and what can be done to secure them.

The front door

Although most break-ins are, as we have seen, at the rear and through a window, some are made through front doors if they look likely to yield without too much trouble.

The average lock used on most front doors is one of the easiest things to open by even a semi-skilled thief. It can be sprung open by means of a thin piece of plastic such as a credit card pushed against the latch between the door and frame – an alternative criminal use of the credit card! If the door has a glass or thin wooden panel near the lock, it can be broken and a hand inserted to turn the lock from the inside (Figure 1).

Many locks have the staple (the part on the frame that the bolt engages with) fixed by short screws into the wood. A shoulder charge or levering with a jemmy can easily loosen these and force them out.

Any such conventional locks, or *springlatches* as they are more accurately termed, should be immediately replaced with a *deadlock*. This is a lock with a bolt that cannot be retracted without a key, that is one that you cannot slam shut but must be locked with a key when leaving. It cannot be sprung open with a card; nor can it be opened from the inside without the key, thus preventing the front door being used as an exit route.

Most springlatches have a two-lever mechanism and, if other means of opening them fail, can be picked without too much trouble. Security deadlocks have a minimum of five levers; some have up to ten levers

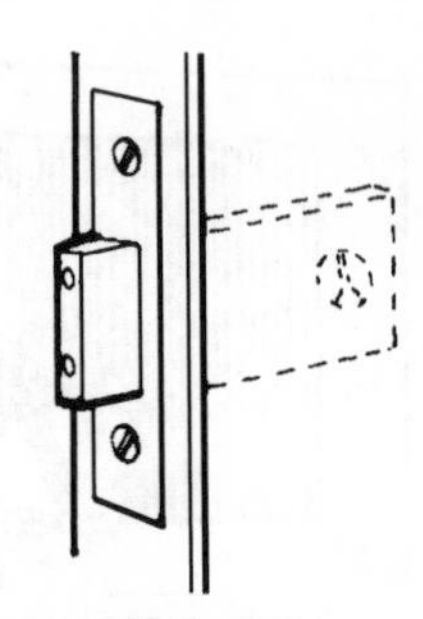

Figure 2 *Mortice deadlock. This should be fitted only when the door is thick enough to avoid weakening it when wood is excavated.*

and are virtually impossible to pick. Another desirable feature is the provision of steel rollers set inside the bolt: any attempt to saw through with a hack-saw is thus futile.

It is generally held that a mortice lock, i.e. one that is fitted in the door rather than screwed to its surface, offers greatest security (Figure 2). This is because it cannot be removed from the inside by an intruder who has gained access at some other point and wishes to establish an exit route; nor can it be burst off the door by force.

However, this needs qualification. Wood has to be removed from the door to make a cavity in order to fit the lock. Thus the door is weakened at that point. If the door is thin, as many modern ones are, there may only be a thin shell of wood enclosing the lock. The application of force could splinter the wood away, rendering the lock useless. In addition, the staple is let into the door frame, and the same weakness could exist there if the frame is insubstantial. So rather than improving security, a mortice lock could considerably reduce it.

Some modern mortice locks are made especially thin to overcome this problem, and less wood has to be excavated. Even so, for a thin door the value of a mortice lock is dubious. It could be added that a thin door in itself is a security hazard and should really be replaced by something more substantial.

If there is some doubt about the matter it would be better to fit a security surface lock such as a *deadlatch*, which offers better protection (Figure 3). Bolts from the keyplate at the front of the door pass right through the door to the lock on the back, and woodscrews enter the door sideways through the side flange in some models. The lock is thus impossible to remove from the outside or inside while the door is closed. As it exerts a clamping effect on the door, it strengthens rather than weakens it. An attempt at forcing an entry is thus unlikely to succeed. The staple is also secured by sideways screws which are concealed when the door is closed and will resist a considerable amount of force.

The type of deadlatch illustrated does not need to be locked shut

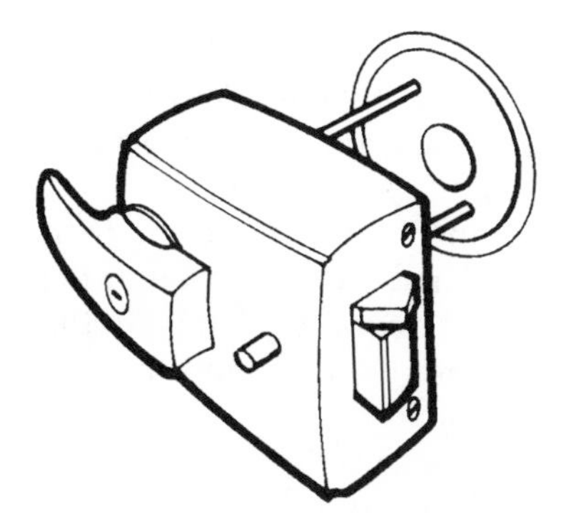

Figure 3 *Surface deadlatch, high-security type. This is fixed by screws at side and bolts through door from front keyplate, and can be double-locked to prevent opening from inside without the key. This type can be fitted in place of many ordinary springlatches without modifying or cutting the door.*

but can be slammed just like a springlatch, and so is more convenient when leaving. However, the shape of the latch and its action make it very difficult to spring open with a card. Another feature is the double-locking facility. The lock has a keyhole on the inside as well as the outside. When the inside is locked the handlever is immobilized so that it cannot be operated to open from the inside. The surface deadlatch is thus really preferable to the mortice lock, although the latter has the reputation of being more secure.

It should be noted that the key of a mortice lock should always be left in the lock at night, and a deadlatch should not a double locked at night. This is in case of fire. Intruder security measures should never jeopardize a fire escape route; better be burgled than burnt!

Often expensive locks are installed on a stout door, but the door frame is thin and so forms the weak link. The door may remain intact after a hefty charge, but the frame may splinter, so rendering all the security devices useless. The frame must therefore be critically examined.

One trick that has been used by burglars is to spring the door frame apart with a car jack at the point where the lock is fitted. Often it can be bowed sufficiently to disengage the lock bolt from the staple or rebate plate. To minimize this possibility the frame should have solid support at the sides. Weak materials such as plaster should be excavated and replaced with concrete on both sides of the door, especially near the locks, and any gaps should be filled with the same material. The rebate itself should be not less than $\frac{3}{4}$ in (19 mm) thick.

Further security can be achieved by having two locks spaced well apart. This adds to the inconvenience of locking and unlocking and means an extra key; it is one of those decisions that has to be made as to the degree of security considered necessary. In high-crime areas or if valuables are kept in the house, it would be advisable.

Where very high security is required, even a stout wooden door may be insufficient. Battering rams have been used to break down doors. A wooden door can be reinforced by covering the outside surface with mild steel of 16 gauge or thicker. The edges should be turned over and secured to all four door edges by rows of countersunk wood screws.

In addition, coach bolts should be fitted at intervals of not more than 9 in (22.5 mm), with the heads on the outside, and passing through the stiles and rails (the main vertical and horizontal door members). On the inside, large washers should be fitted underneath the nuts, and the bolt ends should be burred over the nuts.

A good quality mortice deadlock of at least five levers should be fitted, together with mortice lockable bolts if the door is not to be used as the final exit. If it is, then at least one other similar lock should be fitted. The increased weight will require an extra pair of hinges, and dog bolts (described later) should also be fitted on the hinge side.

This will give an extremely attack-resistant door, but do not overlook the frame, which could now be the weak link. As before, it must be well supported by concrete at the sides. In addition, a strip of angle iron screwed to the frame at the opening side will prevent the insertion of a jemmy to force the door and frame apart. It need hardly be added that such a measure is not usually necessary for a domestic front door! It has been described for cases where special security may be required.

Telephone wires

With the smash-in described earlier, or abduction attempts, the telephone could be vital. However, it is soon neutralized by cutting the wires, which is frequently done. If the wires enter from outside, as most do, some consideration should be given to protecting them. Just how this is done will depend on the way the cable is run down the wall from its anchoring point on the eaves, and the type of wall surface. For flat surfaces, it could be easily covered by galvanized metal capping such as used for mains cables, fixed to the wall. Rough stone surfaces, being uneven, would be more difficult. Alternatively, if the down drop was over a front garden, it could be hidden by a bush, preferably a rose which would present a thorny problem to any would-be tamperers!

Rear entrances

As we have seen, many owners make the mistake of fortifying the front of their premises to a high state of impenetrability but have only the most rudimentary protection at the rear. This is in spite of the fact that a break-in at the back of the premises is far more likely. Burglars are well

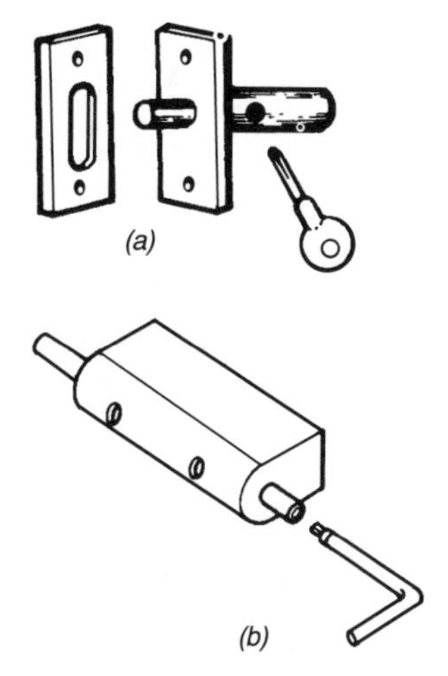

Figure 4 *(a) Mortice rack bolt, with key. This needs only a hole drilled in door or window. The window type is shorter. Another smaller hole is required to insert the key. (b) Surface rack bolt. A key is inserted in the end to operate. All fixing screws are concealed. This type is used where a door or window has insufficient wood for mortice type.*

aware of this quirk, which suits them well because they much prefer to enter at the rear. It is usually less public than the front, so they stand far less chance of being observed.

Rear doors should be of substantial construction and fitted with deadlocks plus lockable bolts top and bottom. These should be secured at all times, even when the premises are occupied, unless access is needed for loading or other essential purposes.

To prevent use as an exit route and for removal of bulky items, the rear door should be secured so that it cannot be opened from the inside without a key. Often, simple draw bolts are fitted and keys left in the locks in the belief that their only function is to keep burglars out.

The lockable bolts mentioned are commonly known as *rack bolts*, and they can be either mortice or surface fitting (Figure 4). A common key fits any bolt of the same type and so enables a number of rack bolts to be used without the inconvenience of needing a key for each. This means that they are not by themselves high-security devices, and should always be used in conjunction with a deadlock. However, the intruder would only encounter them when he is inside and it is unlikely that he would have a suitable key with him. The greater the number of devices securing a door the stronger it is, and the harder and more trouble it is to break open.

Mortice rack bolts require only a small hole in the wood to receive them, so there is little weakening of the door. A version designed for windows is shorter than the door type so that it can be accommodated in a narrow window frame. A second smaller hole intersecting the first at right angles is required for the key.

Surface rack bolts are fitted when it is not possible to use a mortice

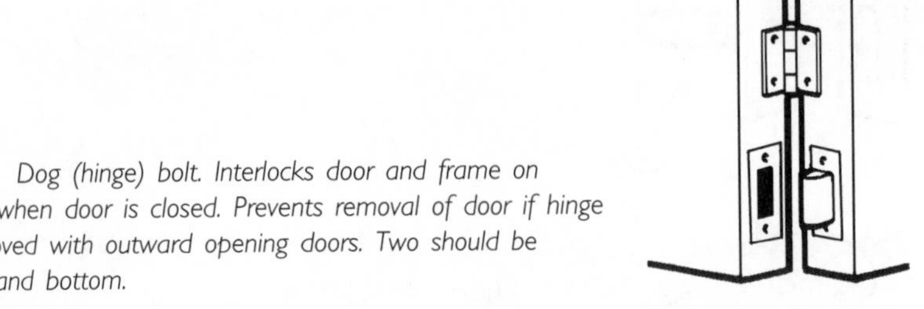

Figure 5 *Dog (hinge) bolt. Interlocks door and frame on hinge side when door is closed. Prevents removal of door if hinge pin is removed with outward opening doors. Two should be fitted, top and bottom.*

bolt. All the screws are concealed once the device is fitted, and it has the advantage that it can be bolted manually without a key, although of course the key is necessary to draw it.

Most doors swing inwards, but any exterior door opening outwards is vulnerable to attack to the hinges, as the hinge pin is exposed on the opening side. It is not too difficult to remove the pin, whereupon the door can be simply lifted away. Fire doors and those fitted to some outbuildings are usually outward opening, so these constitute a major security hazard.

The solution is quite simple, and takes the form of what are known as hinge or dog bolts (Figure 5). These consist of a recess plate and engaging lug which are fitted to the frame and door, respectively, on the hinge side. A set should be fitted to both the top and the bottom of the door. When the door is closed, the lug engages with the recess, so that the door cannot be lifted out if the hinge pin is removed. Even with inward opening doors, dog bolts can be fitted to reinforce the hinges against a forced entry, especially if there is some doubt as to the strength of the hinges. The beauty of them is that once fitted they need no further attention, and as they engage automatically they do not create inconvenience in use.

Windows

The majority of all entries are made through windows, as these are the most vulnerable points and are usually the most neglected. Often windows are left open, thereby offering a clear welcome sign to the intruder. Ground-floor windows are his first choice, but first-floor windows and fanlights are by no means out of bounds to him. Remember, he will most likely be young, agile and thin, so climbing buildings and entering narrow spaces will be no problem for him. Do not think that a window is too small: entries have been known through gaps of only 8 in (200 mm)!

Few burglars will attempt to get through a broken pane of glass. There is too great a risk of being badly cut unless all the broken glass is removed from the window frame, and this would take too long as well as itself being risky. However, some have tried it and been seriously injured in doing so.

Several methods are commonly used to open a window and gain entrance. One of these is by manipulation of the catch from outside. With a sash window, this can often be done by sliding the ubiquitous credit card between the frames and slipping the catch back. Another method, where a small fly window has been left open, is to drop a looped cord inside and engage the main window handle which then can be easily lifted by pulling on the cord.

A common one is to break the window and insert a gloved hand to open the catch. In a refinement of this method, a few strips of tape are first stuck across the window to stop the pieces falling and making a noise.

Double glazing is undoubtedly a deterrent as there are two panes of glass to penetrate instead of one. Furthermore, it is not easy to break a double-glazed sealed unit as a blow to one pane is cushioned by the air trapped between the panes; the action is rather like trying to crack a nut by hitting it when it is supported on a cushion. However, double glazing is not totally burglar proof as is often claimed. There is a way of breaking the glass, which we will not describe here in case this book falls into the wrong hands.

Weathered and broken putty is a major security hazard as it can be easily chipped away and the window glass lifted out. Then the catch can be released and the empty frame opened; this method is often used. Another is to prise the window away from the frame with a jemmy.

As all these modes of entry except pane removal depend on the window being opened, security can be achieved by preventing it from opening, providing the frame and puttywork are sound. There are a number of ways of doing this.

The short mortice rack bolt already described is one. This is ideal for the sash window, in which two bolts should be used, one on each side of the frame. These can be fitted so that the window can be locked when opened a few inches to allow ventilation.

The conventional window fastening for a hinged window consists of a pivoted handle and catch which can be easily released if the window is broken. These can be obtained with a built-in lock (Figure 6), and are now often supplied with replacement windows. These should always be specified for new windows, and it may be possible to fit one in place of an existing non-locking catch. Those that are welded or riveted to a metal frame and so not easily removable can be secured with a locking arm that holds the catch in the closed position (Figure 7). A similar device can be obtained that secures the window stays that are commonly used for fanlights.

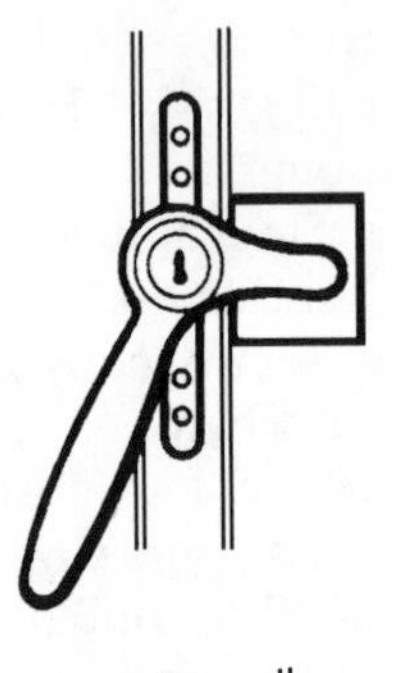

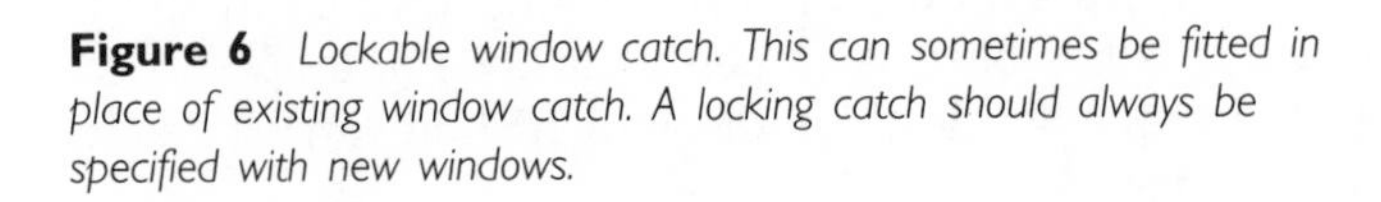

Figure 6 *Lockable window catch. This can sometimes be fitted in place of existing window catch. A locking catch should always be specified with new windows.*

Figure 7 *Window catch lock for fitting to existing catches. The arm swings up under the catch and is locked in place.*

In some homes there are windows that are never opened, such as those in unused spare bedrooms, entrance halls and stair landings. The best security for these windows is to permanently seal them by simply screwing long woodscrews at an angle through the window to its frame. Sash windows can be screwed to each other. This of course is reversible, and if in the future the window needs to be opened it can be easily unscrewed.

In some cases an intruder has used a glass cutter to cut around a pane close to the frame and thereby to gain access to the catch or, if the opening is large enough, to climb through. He may cover the cut edges with several layers of sticky tape to prevent injury as he does so. This calls for some skill and is unlikely to be done by the average opportunist teenager. But it may be attempted by the more experienced burglar who is after something he believes is worth the effort; this is likely to be the type who clears houses when the occupants are away. It is more likely too with a specially vulnerable window, i.e. one that is on the ground floor and out of general view, and which for some reason cannot be protected by the devices we have described.

One effective counter-measure is to fit an iron grille inside the window. Now this idea may seem unpalatable, spoiling the look of the window and giving the impression of living in a fortress, but it need not. A very acceptable version of the grille is scrolled ironwork as shown in

Figure 8 *Example of scrollwork that can both protect and enhance a vulnerable window.*

Figure 8. If painted white or a pastel colour to match the room decor it can actually enhance the room's appearance.

Most towns have firms that will make this sort of thing to measure. The inside measurements should be taken very accurately, measuring all four sides, then subtracting $\frac{1}{16}$ in (2 mm) to allow for warped or non-parallel surfaces and painting.

The frame must be made to these measurements, with holes drilled for the fixing screws, then filled with scrollwork which is welded to the frame and to all points where it contacts, making a very rigid and impenetrable barrier. The whole should be painted with a rust-inhibiting paint before fitting, and long screws used to secure it to the window frame. It is visible from the outside and so deters any attempt to break in.

It should be mentioned here that some DIY aluminium windows that were popular a few years ago are a considerable security hazard, as they are screwed into the wooden surround by woodscrews which are easily accessible from the outside. A few minutes work with a screwdriver and the window can be lifted completely out. It is best to replace any of these with properly made units. However, the windows can be made more secure by removing each screw and dipping the thread into liquid glue before replacing it. When the glue is set they will then be almost impossible to remove.

Louvred windows are also vulnerable as the louvres can usually be removed from the outside. These should be replaced, as little can be done to improve their security.

A further protection for vulnerable ground-floor windows could be the growing of rose bushes beneath them. When up to window-sill height, which should not take too long for vigorous specimens, they would greatly restrict access for any nefarious purposes, and would also present a pleasant outlook during the summer.

Fire safety

While it is necessary to secure windows from the possibility of intrusion, this should not be done to the extent that all escape routes are sealed if a fire should break out. First-floor windows often are the only means of escape for persons trapped in the upper storeys of a building when fire engulfs the lower floor or the staircase. For all windows to be screwed fast or to have non-removable grilles could create a death trap.

Usually, the front first-floor windows, that is those facing the street, are less likely to be entered by an intruder because of their height and the risk of being observed. These are also the ones most easily reached by fire escape ladders. At only a slight extra security risk, then, these should be easily openable from the inside. They can of course be wired with alarm detectors.

Security coatings

Access to upper storeys and flat roofs is often easily achieved by climbing drainpipes, posts and similar objects, while fencing and boundary walls are all vulnerable to scaling.

An effective preventive is a special anti-climb paint such as manufactured by Camrex Special Coatings. It is applied as a one-coat layer $\frac{1}{10}$ in (2.5 mm) thick over the surface to be protected. The treated surface is as difficult to climb as a greasy pole. It looks dry but stays soft and slippery for many years whatever the weather, and does not drip or sag in hot conditions. Any attempt to climb it covers the hands, shoes and clothes in a clinging sticky mess. No doubt parents of youthful offenders would ensure that the attempt was not repeated!

It contains an easily detected chemical that assists identification of suspects, yet it is non-toxic and does not permanently harm clothing. There is thus no risk of legal come-backs. But to avoid innocent inadvertent contact it is recommended that the coating is not applied to surfaces below 7 ft (2 m).

Graffiti are another problem that can cause heavy expense due to the frequent need for redecoration of exterior walls, fencing etc. The same firm makes a range of anti-vandal coatings which are said to be virtually impossible to mark with ball-point pens, lipsticks, crayons and felt-tip pens. The surfaces are impervious to aerosol paint sprays which can be easily removed with stain remover. Various textures are available from heavy to smooth: some are coloured or random patterned containing multi-coloured flakes, while others are clear to allow the original stone or brickwork to be seen.

Outbuildings

Outbuildings and sheds may contain little of value and so not warrant any special security. It is important, though, to secure items such as ladders, crowbars, garden spades and the like, as these could be used to gain entry into the main building. Padlocks are usually used to protect such premises but, like door locks, they differ in the degree of security offered. It is little use having a high-security padlock with a fitting that can be simply unscrewed from the door, or from which the hinge pin can be punched out. Locking bars should be fitted that conceal all screws when closed and have countersunk hinge pins.

As well as being used to gain entry, ladders are a popular item for theft, according to police reports. It seems they are used by builders and window-cleaners in the 'black economy'. They should never be stored outdoors, but should be locked away inside and chained and padlocked to some structural member. Chains should have welded links, otherwise the link joins can be readily pulled apart, and should be substantial. Furthermore, your name and address should be painted prominently along both sides of the ladder.

Why have an alarm?

If the measures described in this chapter are carried out, a major step will have been taken to guard your home from unwelcome visitors. This is not to say that an alarm system is now superfluous. All good generals have a second line of defence, and that is what an alarm system is. Its obvious presence will deter many would-be burglars, and, should any get through the physical defences, it will call attention, distract and unnerve the intruder, and possibly thwart personal attacks and abductions.

Really, the physical security and the alarm should be integrated into a single effective system of defence. This will be discussed in Chapter 8, 'Planning the system'.

2 Alarm system requirements

Our first step in considering an alarm system is to define precisely what is required of it. The principal function is usually considered to be to warn or inform others that an intrusion has taken place in the premises concerned. This really is of secondary importance, especially, as is often the case, little notice is taken of an alarm owing to the prevalence of false alarms.

Rather, the prime function is to deter. This word is derived from the Latin *terrere*, to frighten, and often the very sight of an alarm bell outside the building and other signs of electronic detection are sufficient to warn off the would-be intruder. Even though alarm bells or sirens are often ignored, they are bound to attract some attention, and the last thing most intruders want is publicity. Their aim is to get in and get out unobserved and unrecognized. If there is too great a chance of being detected, they will look elsewhere for a less risky place – and there are usually plenty of them!

In most cases entry is attempted at the rear of the premises where security is often poor and there is less chance of being observed. If the alarm system is set off, it takes stronger nerve than that of most intruders to enter with the alarm sounding. The immediate reaction is one of panic and an overwhelming desire to get away quickly. Usually the louder and more strident the noise, the greater the panic it produces.

This of course is just what is wanted. Some control units allow the audible alarm to be delayed while immediately signalling a break-in to the police or monitoring service via a telephone line. The object is to catch the culprit red-handed, but the wisdom of this is dubious. It is far better to scare off the intruder than to allow him to enter in the hope that he may be caught. He may still escape before the police arrive, and could get away with valuables or do considerable damage in the meantime. A good rule is: *prevention is better than apprehension.*

To accomplish this effectively, the alarm should be loud and clearly heard wherever an entrance is possible. This means that a sounding device should be provided at the rear of the premises as well as at the front. It is not often done, but is well worth the small extra cost. In addition, an internal bell is an excellent deterrent. It would take very strong nerve to actually enter premises with an alarm bell sounding inside.

Reliability

A vital factor is reliability. If the system is out of action for even a short while, that could be the very time the intruder strikes. Reliability is dependent firstly on the control unit, its design, construction and quality of components, and secondly on the installation, its wiring and ancillary equipment including the sounding device.

The reliability of the control unit cannot be assessed by a potential user. It would require experience of a large number of the same model, something that only the large installation companies would have. If they have a poor record of reliability for any particular model it is unlikely they would continue to stock or supply it. A model that has been on the market for some while and is still readily available from several firms is thus likely to be a better buy than a new one, however attractive its features may be.

Actually, all electronic components are liable to failure. The failure rate per thousand samples, plus the mean time to expected failure, are assessed by exhaustive testing and are specified by component manufacturers. There is a British Standard as well as IEC standards that define methods of expressing failure rates.

This shows that 100 per cent reliability for components is unattainable, and any claims for such should be treated with the utmost scepticism. It follows from this that the more components a control unit has and the more complex it is, the more likely statistically it is to fail. Complex units having many features are now offered at quite reasonable prices compared with what they were at one time. This is partly due to the use of dedicated (specially designed) silicon chips which carry out most of the functions. However, a host of features you will never use could be obtained at the cost of higher liability to failure.

This is not to say that the present generation of alarm units is basically unreliable, only that the chances of failure are greater with a more complex unit. Having made this point, it should be said that the most likely cause of a fault is in the installation – the wiring, the sensors or the use of unsuitable sensors.

False alarms

One of the biggest problems with alarm systems is that of the false alarm. In the Metropolitan Police area no less than 98 per cent of the call-outs are for false alarms. Understandably the police see this as a considerable waste of their time and resources. Some forces have laid down conditions as to how far they will respond: for example, more than five false alarms in any month and the owner is warned; after three months of nuisance calls there is no further police response to an alarm.

Apart from police involvement, the goodwill of neighbours will inevitably be jeopardized and strained by frequent false alarms, to say nothing of the trauma experienced by the householder who is frequently disturbed from his bed, or called home by the police during the day. It is the high incidence of false alarms that has resulted in general scepticism and the common ignoring of alarms when they sound.

A trick that has been used by some burglars is to try to deliberately set off the alarm in some way without making an obvious entry. The householder, thinking it is another false alarm, and not wishing to be disturbed again, switches the system off. The National Supervisory Council for Intruder Alarms (NSCIA) code of practice actually recommends not switching the system on again after a false alarm. So when the fuss has died down the intruder breaks in, secure in the knowledge that he will not be disturbed.

This ruse can only work if the system is prone to false alarms and can be triggered from the outside or without causing obvious damage. Illustrating how over-sensitive some systems are is the fact that one night in October 1987 a security firm in the London area logged 3000 calls in a period it would normally receive 8 to 12. The police computer handling incoming calls broke down under the strain. The reason? That was the night of the hurricane which swept the country, and the alarms were due entirely to the high winds and their effects.

It can be seen, then, that the false alarm must be avoided above all else, but in a way that does not compromise security. It can be done, as there are a large number of effective installations that rarely if ever experience one. Human error is sometimes to blame, and little can be done about that except to stress extreme care by everyone concerned. Most of the trouble lies in the installation of the system or the use of unsuitable sensors. This aspect will therefore be stressed in Chapter 9, 'Installing the system'.

Convenience and cost

One factor that is sometimes overlooked is the balance between the level of appropriate security and convenience. This was mentioned in the previous chapter in connection with physical security and it is true also of alarm systems. Some set-ups can be nerve-racking to live with: a complex setting and desetting routine has to be remembered; there are too many and too sensitive sensors that are liable to be triggered at any moment; and children and pets have to be watched so that they do not set off the alarms.

Another factor is cost. It is possible to fortify premises to a standard that would do justice to a bank, at considerable cost to install and, in the case of telephone monitoring systems, a recurring sum to check and maintain. This may indeed be desirable for large homes housing valuable collections or art treasures, but is it worthwhile for the average home?

Security costs money, so buying more security than you really need is money wasted, although it is always better to err on the safe side. A really determined professional crook will gain access almost anywhere and go to infinite pains to do so. He can defeat even the best security systems. However, such individuals are fortunately rare and only a small percentage of the total number of intruders. Most are opportunist thieves looking for an easy 'job', many being local youths or even children.

Quite a high degree of security can be achieved, sufficient to defeat the efforts of such individuals, at moderate cost and with only slight inconvenience. However, the mistake must be avoided of cutting costs bv making part of the premises secure while neglecting other parts. The old adage about the strength of a chain being that of its weakest link is very true here. A weak point, whether in physical security or in the alarm system, is sure to be spotted by the crook. Money spent on other areas of security will then be like stopping one hole in a colander.

In the case of security installations, a question to be considered is whether to call in a professional firm of installers or to attempt a DIY job. With the average home, a DIY installation will be perfectly satisfactory, providing it is well planned, the householder has a good working knowledge of the various bits and pieces and their capabilities and limitations, and it is installed to a reasonable standard. Guidance on all these aspects will be found in succeeding chapters. All that is then required are some basic DIY skills such as drilling, screwdriving, hammering staples, connecting wires to terminals etc.

Homes with special security requirements such as those housing valuable objects may be subject to the attentions of the professional thief, and so a professional eye may be needed to assess the risks and plan the system, and skilled tradesmen may be required to make the installation. Even so, with the help here given, there is no reason why you could not install an effective system yourself.

There is one other consideration though. Some insurance companies give discounts for alarm installations but only if carried out by an approved contractor. Whether the discount over several years is worth the difference in cost between a DIY and a professional installation is something that would have to be investigated. Some companies may even refuse to insure very costly items unless the system has been professionally installed. In such case you have no option but to have one. However, with the information provided in this book, you will be able to discuss the matter knowledgeably, and make an informed choice between the various options that may be presented by different firms.

Having established the essential system requirements, we can now go on to take a look at the basic alarm circuit and its elements.

3 Basic alarm systems

The alarm system can be broken down into four basic essentials. First there is the sounding device which in most cases is a bell but sometimes is a siren. Second there is a power supply which can be derived from the mains with a battery back-up, or in a smaller system can be just a battery. Next there are switches which are activated by the intruder. These can take many forms and so are more usually described as sensors. Finally there is a master control which switches the whole circuit on or off, or selects test and other modes of operation.

The basic circuit

In Figure 9 we have the simplest possible alarm system containing these four elements. It is similar to an ordinary door bell circuit except that several switches are connected in parallel across each other. There is also a master switch. If any one of the sensor switches is closed the alarm bell will ring. There is no limit to the number that can be wired into the circuit, and they can be of different types, such as door switches that operate when a door is opened, or pressure pads under carpeting that close when trodden on.

As it stands there is a very serious limitation to this circuit. If the bell rings when a door is opened it can be stopped simply by closing the door again. Or, if the alarm is activated by stepping on the carpet, it will stop as soon as the pressure is released by stepping off. Obviously such an arrangement is of little use as a security device.

What is needed is a means whereby the alarm, when once started, latches on and continues to sound irrespective of what subsequently happens to the sensor switches. Only throwing the master switch can then silence it. A latching arrangement is therefore an essential part of all intruder alarm systems.

Before the advent of semiconductors, the method which was universally used and still has a lot to commend it is the *latching relay*. This is a switch that is magnetically operated when an electric current is passed through a coil of wire. When the current ceases, the switch contacts spring back to their former position.

A simple arrangement is shown in Figure 10. The coil is connected across the bell and the relay switch across the sensors. When a sensor switch is closed, current flows from the battery through the bell and also

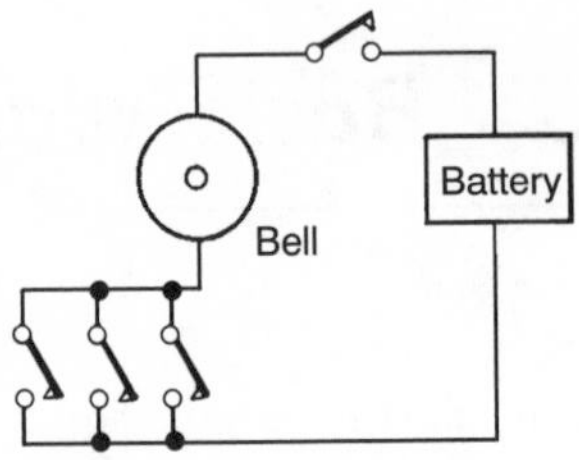

Figure 9 *Basic alarm circuit consisting of bell, battery, control switch and parallel sensors.*

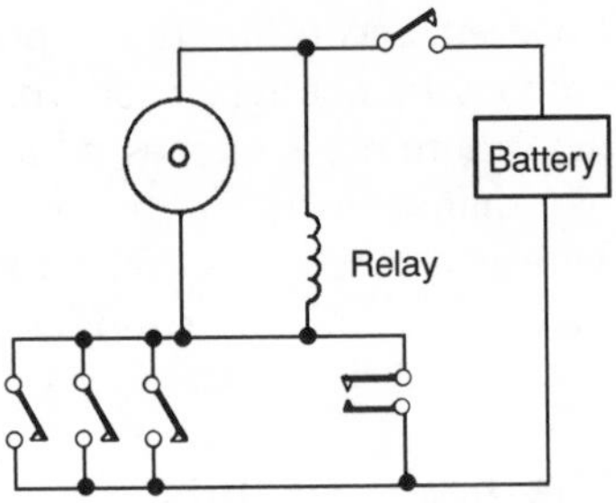

Figure 10 *Using a relay to latch the circuit on. When a sensor closes, current flows through the bell and relay coil. The relay switch closes, short-circuiting the sensor circuit, so maintaining the alarm even if the original sensor is then opened.*

the relay coil. This closes the relay switch which thereby completes the circuit and keeps the current flowing through the bell and relay coil irrespective of whether any sensor is opened or closed. The circuit is thus latched in the 'on' state. When the master switch is opened, the relay current ceases and its switch resets.

Closed loops

The circuit shown in Figure 10 is what is known as an 'open' circuit, that is, all the sensor switches are normally in the open position, and when activated they close. This arrangement has certain drawbacks. It is vulnerable to tampering, to accidental breakage of a wire to one of the sensors, or to damage to the sensor itself. If this happened, that sensor would be inoperative and part of the premises left unprotected.

There would be no means of knowing this except by periodic testing of all the sensors. Such testing could only be done by actually operating each switch in turn, which would be very inconvenient if the alarm was sounded each time. Even if a test circuit was devised to prevent this, it would still be a laborious and time consuming task to check each sensor individually, so the premises could be vulnerable for a long time between tests.

The alternative which avoids these difficulties is the *closed loop*, a basic circuit of which is shown in Figure 11. Instead of the sensors being

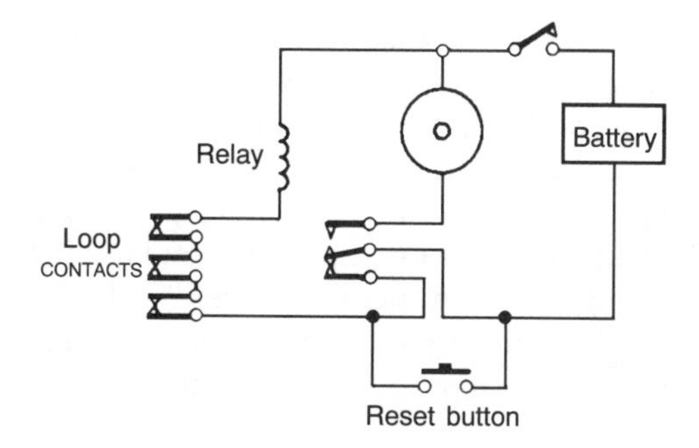

Figure 11 *Relay latching circuit with closed loop. Coil is continuously energized, holding the switch open. If a sensor opens, the relay is de-energized and the switch closes to sound the bell.*

normally open they are normally closed and are connected in a series loop as shown instead of being wired in parallel.

The control unit continually circulates a small current around the loop when in the 'on guard' condition. If any of the sensors are actuated the contacts open, thus open-circuiting the entire loop. Any break in the loop stops the current which is interpreted as an alarm situation and so triggers the sounding circuit.

In the case of the circuit in Figure 11, the relay is continuously energized by the current passing through the loop. The associated switch is known as a single-pole changeover, and has three contacts. One of these, which is common, connects with one contact when the relay is energized, and the other when it is not.

When the relay is energized, current passes through the closed contacts and from there through the loop and relay coil. Any break in this circuit, caused by a loop sensor switch opening, de-energizes the relay, thereby opening the first pair of contacts and closing the second. This pair actuates the bell circuit. As the first pair is now opened, any closure of the loop cannot re-energize the relay and stop the alarm. The circuit is thus latched in the alarm position.

To reset the circuit the reset button is momentarily depressed. This is connected across the first pair of relay contacts and completes the loop circuit, so energizing the relay. Once it is energized, the first contacts are again held in the closed position, thus continuing the loop and energizing current.

If non-electrical readers found those descriptions a bit obscure, never mind; the relay is not often used today anyway, its place being taken by semiconductors known as thyristors (which are even more obscure). The important things to know are: that an alarm circuit must be latched on as soon as it is activated; that an open type of circuit in which all the sensor switches are off has the disadvantage that individual sensors can be inoperative without anyone being aware of it; and that a closed loop has all its switches on and has a continuously circulating current.

The advantage of the closed loop is that if any sensor or any part of its wiring becomes open-circuit due to damage, no current will flow and the alarm will be triggered as soon as the system is switched on. It is thus never left unknowingly in a vulnerable state. As this in itself would be somewhat inconvenient, most control units have a test facility so that a warning light comes on instead of the bell sounding if a fault is present at switch-on.

The snag with this circuit is that the relay current is flowing all the time the circuit is on guard, or 'armed' in the terminology of intruder alarms. The relays that were originally used took a heavy current, and even large-capacity batteries had a short life. More recent relay types were far more frugal in their current demands but still needed a fair amount, enough to drain a battery after a few days of continuous operation.

This may seem unimportant when the supply is obtained from the mains, but is not when battery back-up life is considered. If a mains fault occurs at the start of a long holiday period when the premises are unoccupied for a week or more, how long will the battery standby last? This is obviously an important factor.

The simple relay circuits originally used all but disappeared when semiconductors came on the scene. Either a transistor was used to control the relay or a thyristor was employed which dispensed with the relay altogether. In either case only a small loop current was needed to keep the device turned off, which when removed turned it on, thereby activating the relay or, in the case of the thyristor, the bell directly.

Tamper-proofing

A simple two-wire loop is quite adequate for domestic alarm systems where those having access to the premises when the alarm is off are members of the owner's family or persons known to him. In the case of business premises open to casual visitors or to the public, tampering is a possibility as a prelude to a later break-in.

To counter this, most alarm systems have a tamper-proof loop, also known as a 24-hour loop. Although tamper-proofing is not usually necessary for home alarms, a 24-hour loop or similar provision is required for the panic button which will be described later.

If the loop wiring has been cut during business hours by an intending intruder, it becomes evident when the system is switched on, as we have already seen. But this probably will be when the premises are being closed at night, a very inconvenient time. It is far better to receive a warning when the actual damage is inflicted so that immediate steps can be taken to repair it, and possibly discover the culprit.

To protect the loop and give immediate warning of any tampering, a four-wire system is commonly used in business installations. One pair is the normal loop having all the switches connected in series with it, and the other is a loop which connects to blank terminals on each sensor which are just used as connecting points, or as straight-through links.

Current is passed around the second protection loop continuously for 24 hours a day whether the system is switched on or off. As none passes through the actual sensor switches it is not affected by the normal comings and goings. If the wire is cut, though, the loop is broken and an alarm is immediately sounded. This need not be the main system bells or sounders which could alarm genuine customers on the premises, but a smaller device in the supervisor's or security guard's office. The protection loop is usually non-latching because unlike the sensor loop there is no need to ensure it stays in the 'on' condition: the damage will not repair itself!

The anti-tamper loop can be connected to any vulnerable part of the installation. The bell box can have a microswitch operated by a spring-loaded plunger set under and held down by the cover. If the cover is removed, the switch is released and the protection loop is open-circuited, so triggering the tamper alarm. A similar arrangement operates in the control box and some space protection sensors such as infrared, ultrasonic and microwave detectors.

Added security

This is for those seeking maximum security for an alarm system such as business premises housing very valuable stock, or homes with art treasures likely to attract the professional thief. If yours is an ordinary home you can skip the next two sections.

Where extra security is required, even the normal four-wire system incorporating a 24-hour protection loop could still be vulnerable. It is possible for tamperers to peel back the outer insulation and short-circuit the detection loop wires with a pin, thus disabling the associated sensor switch or switches. The detection loop pair would have to be identified, but short-circuiting both pairs could be easily done without triggering the anti-tamper circuit. The problem for the tamperer would be to identify which were pairs out of the four wires, as there is no standard colour code.

Added protection against this could be afforded by means of sensors using single-pole changeover switches having three contacts, wherein a

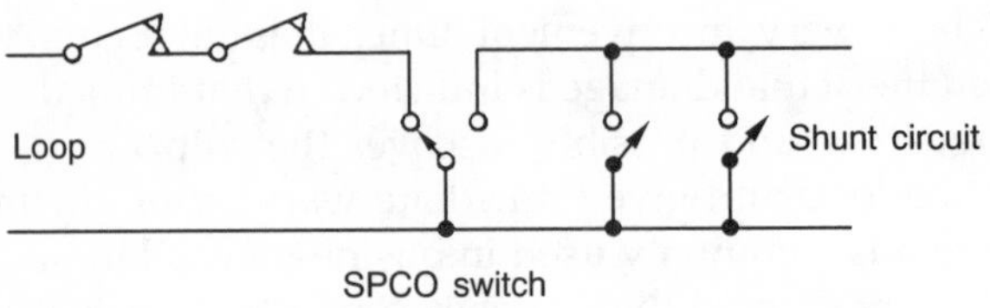

Figure 12 *Single-pole changeover switch connected either in normally closed mode to a loop circuit, or in normally open mode in a shunt circuit.*

common one makes with one contact while it breaks with the other. The pair that is normally closed is wired into a loop in the normal way and the odd one is connected to those of the other switches as in Figure 12. They are then taken to the open-circuit terminals of the control box which are used for pressure mats that are normally open-circuit and cannot be connected in a loop.

Should any of the normally closed loop connections be short-circuited by bridging the wiring, the sensor would still trigger the alarm by activating the open-circuit facility of the system. Furthermore the extra wiring serves to decrease the probability of a tamperer guessing the identity of the wires.

Changeover switch sensors are not often used in spite of their added security, and are not easy to obtain, but there is an alternative.

Dual-purpose loop

The open-circuit facility to which pressure mats are connected requires separate wires back to the control box. So, if a mat and a loop sensor are situated some distance from the box, either a separate run must be made or extra cores included in the loop cable. This can be obviated in systems that have a dual-loop facility (Figure 13) which can take both normally open (pressure mats) and normally closed (door contact) sensors.

It works like this. A terminating resistor is included at the furthest part of the loop and higher-value resistors are connected across each pair of series sensor contacts. Pressure mats are connected across any part of the loop. A continuous current is circulated in the 'day' mode irrespective of whether the contacts are open or closed, because the loop is completed by the sensor shunt resistors. Resistance will vary as doors are opened and shut and so also will the circulating current, but the control unit ignores these variations. If though the loop is broken by tampering or other damage, the current ceases altogether and this immediately triggers

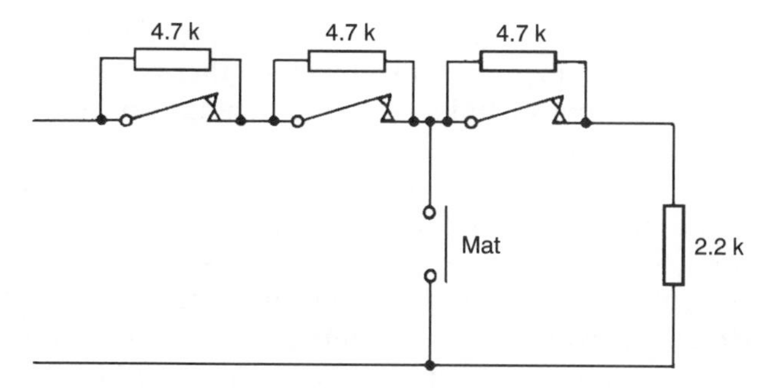

Figure 13 *Dual-purpose loop with end-of-line resistor. Pressure mats can be connected across the circuit, thereby causing a short-circuit when operated. Series switches with shunt resistors can be wired in a loop, producing a high resistance when opened.*

a tamper warning indication. Thus the loop is continually monitored without the need for a separate anti-tamper circuit.

When switched to the 'night' mode, the control unit is now sensitive to current variations, so if a series sensor switch is operated the loop resistance increases and the current falls, or if a pressure mat is actuated the resistance drops and the current increases. Either condition triggers the alarm. Both normally open and normally closed sensors can thereby be mixed.

Another common method is to wire the pressure mats from the detection loop to the protection circuit. With control units that are designed for this mode of connection, any short-circuit appearing between these loops initiates the alarm.

Zoning

With large houses it may be necessary to protect part of a building that is unoccupied when another part is not, whereas at other times all parts are unoccupied and need protection. In small businesses, for example, there may be living accommodation in the same building as the business, such as a flat over a shop.

In such cases it is very useful to be able to switch on only part of the alarm circuit that covers the unoccupied areas. Even in many smaller homes this could be desirable, as intruders have been known to enter the back of the house while the family were watching television at the front.

Systems can be divided into two or more zones, and the control units carry a specification as to the number of zones that can be served. Usually, one zone is activated all the time the system is switched on, while the others can be switched in as required. All sensors serving the first area to be unoccupied are connected to the first zone, while those protecting areas that are still occupied are linked to the other zones.

With single-zone control units, separate zoning can be achieved by simply installing a switch to short-circuit part of the loop. It could be located near the control panel or at some other convenient point. To be secure, it would need to be hidden or disguised if installed remote from the panel, and to be *itself* within the area still protected when it was switched off. This is important as it thus could not be operated by an intruder without activating the remaining sensors. A disadvantage is that it could be forgotten and left off when total protection was needed.

With large premises, zoning offers another major advantage. Most zoned control units have visible indicators to show which zone tripped the alarm. If more than one zone has been activated, many panels indicate which was the first one. This enables a speedy identification of the area of actual or attempted entry. The indication remains even after the bell may be silenced by a timer.

Zone identification with large systems is desirable not only in the case of activation by an intruder, but also if there is a false alarm. The source of these is often difficult to determine, and a zone indication narrows the field considerably.

Panic button

All systems now have the facility of being triggered by a press button, even when they are switched off. This is generally known as a panic button, and serves to summon assistance in an emergency or to scare off an attacker. The buttons are usually of the normally closed type that open when pressed and so can be connected to the 24-hour protection circuit.

Many systems have a separate personal attack (PA) facility. As some PA circuits are non-latching, the panic button itself needs to latch. It also needs to be defeat-proof so that the assailant cannot reset it and thus silence the alarm. This is done by making the button resettable only with a key. Most panic buttons are recessed to prevent accidental operation.

If several panic buttons are to be included in the system they can be connected in series in a loop. This confers the same advantages as the main sensor loop. All can be tested without individual operation by checking loop continuity, and any break in the wiring is readily detected.

Exits and entrances

Having switched the system on, the occupier now has the problem of how to get out of the building! There are many possible methods and all of them have snags. The usual method is to have a special circuit called an exit loop to which sensors on the exit door and any others that may be encountered on the way out are wired. The loop is subject to a time delay that can usually be set at the control box for any time from seconds up to several minutes.

Once the system has been switched on, the delay timer starts running. All sensors are immediately operational except those on the exit loop which remain inactive until the delay time has expired. After this, any actuation of a sensor on the exit circuit starts another timer running, and when this has expired the alarm sounds unless the system has been switched off.

This second timing period is to allow re-entry without sounding the alarm. Although in most cases the same time would be required to enter and switch off the system as would be needed to exit after switching on, the timers are generally (though not always) independently adjustable.

The occupier thus has a certain limited time both to exit and to enter and this could give rise to false alarms should he be unexpectedly delayed. For example, dropping keys, spilling the contents of bags or having problems with children could cause the exit door to be opened after the exit time had expired. This would not immediately sound the alarm but would start the entrance timer. Thus the alarm could sound some while later, possibly after the occupier had left and passed out of earshot.

To reduce this possibility, many systems have an internal buzzer which sounds when the exit timer is running but stops when the time has expired. So as long as the buzzer is sounding and can be heard from the exit door, it is safe to leave. Should it stop before the exit is made, the system must be switched off and switched on again. However, the buzzer could stop a split second before the exit door is fully closed on leaving and the cessation be unnoticed; then the entrance timer would start running and an apparantly inexplicable false alarm would result.

The buzzer usually also sounds during the entrance period. It thus serves as a reminder that the system is still switched on and must be turned off immediately if an alarm is not be sounded.

If the exit and entrance timers are set to too short a period, every exit and entrance is a stressful rush to beat the timer; sooner or later a delay will occur. Intruders rarely try to force an entrance within minutes of the premises being vacated, preferring to wait until the occupant is well clear. Even if they did they would almost certainly trip one of the other internal sensors. So the best course is to set the delays to give ample time to exit and enter; control units having only short maximum exit and entrance times are best avoided.

Another exit arrangement is the *exit set*. With this, when a sensor is activated on the exit circuit by the occupier leaving the premises after switching on, the whole system is thereby set. It is not primed until then, no matter how long the system has been switched on. So any delay *en route* poses no problem. Re-entry is by the usual delay.

An alternative method of exit is to use the regular sensor loop for the exit door but to fit a key-switch that bridges the door sensor in the door. When the door is shut and locked on leaving, the key-switch is set to the open-circuit position, and when entering it is switched to closed-circuit.

This can work quite well with commercial installations but is less successful with domestic ones. The problem with these is that various members of the household enter the main exit door using their keys while the house is occupied and therefore while the alarm system is switched off. If they operate the key-switch out of habit, it may be in the wrong position next time the system is switched on. Furthermore it needs to be open-circuit when the system is switched on at night so that the door is protected, but closed-circuit when switching on during the day to permit exit from the premises. All this can generate much confusion and false alarms.

Business premises are normally unoccupied when locked up and so these problems do not arise. The bridging key-switch is thus a viable alternative to the timed exit. The main risk is from forgetting to operate it, but as other keys have to be used at the same time to operate the locks this is unlikely. Apart from this, the snag is that of yet another key to use and carry around!

Another version of the key-switch, which overcomes most of the problems associated with it, is the lock-switch. This is a high-security mortice lock that has a built-in switch that operates each time the lock is turned.

The lock is fitted in place of the existing one and so no extra key is required, nor can it be forgotten or left in the wrong position. It is probably the ideal solution to the exit and entry problem in certain situations. It could of course be fitted as an extra lock to the present one and, while this means an extra key, it also provides additional physical security.

The switch is usually of the single-pole changeover type which can be used to shunt the door sensor or to remotely set some types of alarm circuit. When it is used as a sensor shunt the system must be switched on before leaving and off after entering in the usual way.

With some alarm systems it is possible to switch the entire system on and off by means of the lock-switch on the exit door. A buzzer fitted near the door sounds continuously to warn of a protected door being left open, or other sensor being left in an actuated condition such as a chair standing on a pressure mat. There is thus no operation of the control panel at all.

This type of arrangement is ideal for premises such as church and public halls, where many parties may be allowed access with a key but not all can be relied on to operate an alarm system and carry out a proper exit routine. This is the method used in the second Sureguard circuit described in Chapter 14.

Wiring to the lock-switch is in this case vulnerable to tampering as the whole system could be disabled by it. It could be protected by a 24-hour anti-tamper loop run right to the lock, and the wiring should bridge the door-to-frame junction with a proper door loop consisting of multi-core flexible wire in a sheath terminated by enclosed junction blocks that have anti-tamper lids. Alternatively, the wiring from the lock to the loop could be sunk in a groove in the woodwork which is then covered with filler and painted over so that there is no observable trace, or it could be covered with metal channelling.

Lock-switch setting, though ideal for this situation, is not the answer in premises where different zones are required to be switched on independently. It is only feasible when the whole system is to be switched on at the same time when the premises are locked up.

The bell circuit

The sounding device can be a bell or a siren, but usually it is a bell for reasons we shall explore in Chapter 7. It is obviously vulnerable, being on the outside of the premises, and if it is silenced the whole system is to no avail.

The first essential then is to position the bell where it cannot be reached other than by a ladder. More than one bell on different sides of the building increases the security as well as making more noise over a wider area. An internal bell is also highly desirable for the unnerving effect on any intruder who may somehow manage to breach the perimeter sensors but sets off one of those inside.

Bells are commonly housed in steel boxes with anti-tamper switches under the lid connected to the 24-hour protection circuit. Any removal will thereby set off the alarm. However, a possibility – which although unusual has happened – is for the box to be filled through the louvres with a foam sealer by bogus workmen or window-cleaners. This takes but a few minutes using a pressure canister and effectively stifles the bell. Later they return for a much muted entry.

A further disadvantage of the bell box is that the sound must escape through narrow louvres and is therefore reduced in volume. In particular the high-pitched bell tones, on which the stridency depends, are curtailed.

A better alternative is to use a weatherproof underdome bell with a close fitting dome, mounted without an enclosing box. Some of these are very difficult to silence mechanically, and they give a high sound output.

Wiring to the outside bell is likewise vulnerable and must be protected throughout its run. It should not be run along an outside wall but come straight through a hole from the inside to the back of the bell. The run back to the control box should be concealed and protected from accidental damage, and if possible buried in plaster.

With business premises or those at special risk, a self-activating bell or a self-activating module installed with a conventional bell will increase the security. This device works with a rechargeable battery included in the bell box. It monitors the bell circuit and if the wire is cut it will sound the bell from the internal battery. Most self-activating circuits have a built-in charging facility that keeps the battery fully charged from a supply derived from the control box. If the alarm sounds without any damage to the wiring, power is drawn from the control unit in the usual way and not the battery. Thus the battery is conserved for emergencies.

4 The control unit

The heart of any security system is the control box. Whatever facilities – or limitations – the system possesses, they are determined by the control unit. Sensors and sounders are for the most part compatible with the majority of control boxes, so there are usually no problems assembling the various parts of a system. The first decision to be made is what facilities are required; then a unit offering these can be sought.

There is a huge selection of control units now available; some are highly complex and fairly bristling with features, others are quite basic. At one extreme are those intended only for domestic use, while at the other are those designed for large factory premises. We will take a look at the facilities common to all units, and those which differ from one model to another. We will also consider some features that are not normally required for domestic systems so that the reader will recognize them if they are offered with any particular control unit.

Cost and complexity

First, a word about cost. Control units have dropped sharply in cost owing to the development of special chips for security applications and to increasing demand. Full-feature units now cost little more or even less than the basic models of a few years ago. Even so there is no point in paying for features you will never use, and, as pointed out in the previous chapter, reliability decreases as complexity increases.

However, it may be that features not thought necessary now could prove useful in the future. For example, a garage or home extension may be built which would need different sensors than those already used, or for which a separate zone would be desirable. So the present and future needs should be carefully considered.

Multi-zoning

The provision of a closed-circuit loop is common to all units as it is the principal mode of detection. Some are simple loops with separate anti-tamper circuits, while others have end-of-line resistors and use resistor-shunted contacts to eliminate the separate anti-tamper circuit. These have already been described.

Units differ in the number of individual zones that can be served. Unless zones are mentioned in the specification it can be assumed that the unit is a single-zone type and is intended for domestic use. Two- and four-zone boxes are common, while some are extendable up to sixteen zones by means of fitting extra printed circuit boards. This should be sufficient for all except the largest factory complex for which some models can be expanded up to 320 zones.

It is sometimes advantageous to have two or more zones. These could be used to serve different parts of a large house, a house and attached garage, or the shop and living accommodation of premises containing both. In the event of an alarm, actual or false, the precise circuit or area responsible can then be quickly identified.

As well as serving different areas, another use is to take different types of sensor. Thus movement detectors or pressure mats can be connected to a different zone than the door and window switches. The former can then be switched off when all the family are home in the evening, while leaving the perimeter sensors on to guard against a break-in at the back or other remote part of the perimeter.

Zoned control units identify the zone which has been tripped, usually by means of an indicator lamp which remains latched on until reset. With some, if more than one zone has been tripped, an indication is given as to which was the first zone to be actuated. This may be done by the appropriate lamp flashing on and off, while those of other tripped zones remain permanently lit.

Annunciators

Large installations having only a single-zone control unit can benefit from the addition of an annunciator. This enables immediate identification of tripped sensors as with a multi-zone unit, but may have no provision to actually control any of the sensors. In one model up to six sensors or groups of sensors can be connected with only four wires (six if an anti-tamper loop is included).

This trick is performed by a technical method known as multiplexing. Different value resistors supplied with the unit are wired in series with each sensor; when one is tripped the device recognizes that particular resistance value and a visible indication is given which latches on until reset.

The advantage is the simplified wiring and part multi-zone operation in identifying tripped sensors, but the disadvantage is that individual areas cannot be controlled as with true multi-zone working. It is not an item likely to be needed in a domestic system.

Normally open-circuit loop

All units accept normally open-circuit sensors such as pressure mats. With many domestic control units, the open-circuit loop is separate from the closed-circuit one.

The practice with industrial systems is to combine them in the one circuit, a common method being to connect pressure mats between the loop and anti-tamper circuit. With control units so designed, when these two circuits are short-circuited together, the alarm is triggered.

With the dual-purpose detection loop described in the previous chapter, the closed loop has an end-of-line resistor. Mats are connected across the loop, thereby short-circuiting the resistor if one is actuated. The advantage is that only one pair of wires is needed back to the control box; the disadvantage is the accommodation of the sensor and final resistors. For domestic systems where long wiring runs are not normal, the two simple separate loops, one open and the other closed, are the most straightforward.

Anti-tamper circuit

The anti-tamper loop, active for 24 hours a day even when the unit is switched off, is found on all business systems except those having the dual-purpose detection loop, in which a current continually circulates around the loop in the 'day' mode. Continuous loop monitoring or an anti-tamper circuit is necessary for these owing to the possibility of tampering during the time that the premises are open to the public.

Some domestic units have the facility, but it is not really necessary as tampering is an unlikely possibility. If there is no provision for a panic button circuit, however, a 24-hour loop will be required to take its place. Most modern units do have a panic button circuit. In larger homes where valuables are kept, an anti-tamper circuit may be advisable as tamperers may gain access during the day by various ruses. Four wires must be run to all sensors when an anti-tamper loop is used. This is no problem as multi-cored cables are available with four, six or more conductors.

Exit facilities

Exit circuits are provided in all systems but they differ in the way they operate. The various methods were discussed in the previous chapter. Most control units provide a closed delay loop that becomes active after the delay has expired. The exit door sensor must be wired separately to this loop. Provision is made to adjust the delay to suit the time required to exit. If an alternative method such as a shunt switch on the exit sensor is preferred, the timed exit loop can be ignored.

Provision for the connection of a buzzer is commonly made with most control units. This sounds during the exit and entry periods and with some models is also used for testing, to indicate a fault or an anti-tamper circuit break. It can be mounted near to the unit and wired to it, but many units have a built-in buzzer. Alternatively, some have a built-in loudspeaker that generates one or more tones. Internal sounders save a little on the installation and are generally more convenient, but they have one snag. Few boxes could withstand a hefty onslaught with a jemmy or sledgehammer, so it is desirable to conceal the control unit as far as possible. But if intruders broke through the exit door, thereby starting the entry timer, an internal buzzer could lead them right to the unit.

A buzzer that is audible at the exit door and the control box, although mounted at some distance from them, would confuse rather than aid intruders. Some units allow the internal buzzer to be disabled and an external one connected. Whether this is done depends on the level of security required. In most cases, the convenience of the internal one overrides the slight chance of it aiding an intruder, but for high-risk situations a separate buzzer is certainly desirable.

Testing

All except the simplest domestic control units have provision for testing the detection circuits but the extent of the tests varies considerably. For the basic test the system is activated but the internal buzzer or an indicator lamp is switched in place of the alarm bell.

If there is a fault, that is a door has been left open and the loop is broken, or any other sensor is actuated, the fact is thereby indicated without the alarm being sounded. In the case of a zoned system, an indicator shows which zone is at fault. The system must then be switched off, the fault investigated and corrected, and a further test made. If there is no fault some units have an 'all clear' indicator which shows that it is in order to switch on.

With some models, the main switch must be taken through a test position before it reaches 'on'. Thus it cannot be switched directly to 'on' without first checking for faults. However, the switch could be turned straight through the test position to 'on' and a fault indication ignored. If this is done with some units, the main sounders will not be activated, so preventing a false alarm, and the internal buzzer will continue to warn of the fault condition.

Other models have no separate test switch position. With these the system automatically goes into a test mode when it is switched on, and any fault is shown by indicators or a buzzer, with the main sounders silenced. In some cases when the circuits are clear, all the fault indicators flash on and then go out, to show the circuits have been checked.

These tests merely check that no closed loop is open-circuit; they do not test individual sensors. Nor in most cases do they test the control unit itself; a 'no fault' indication could be produced because the control circuits failed to detect one and so would likewise fail to respond to an intruder activation. For the door switches most faults would show as an open-circuit loop. This is not the case with space protectors which could fail without warning, or normally open-circuit detectors such as pressure mats.

Although it is not feasible to do so at every switch-on, these sensors should be tested regularly, and this is accomplished by what is known as a 'walk test'. A provision is made in most control units to switch out the main sounders and substitute the buzzer so that each sensor in question can be activated without raising an alarm. Each zone loop should also be checked by opening a door on that circuit. There is not usually any way of testing the 24-hour anti-tamper control circuitry other than by open-circuiting the loop to see if a fault indication appears. With industrial systems all this should be done during regular maintenance visits.

The bell is not usually tested other than by an occasional short actual sounding test. Some control panels have a provision for doing this, and a few permit the sounder to be operated at low level. It is possible to test the bell without sounding it at all by passing a small current through it. This is a feature believed to be unique in the Sureguard design described in Chapter 14.

The self-actuating circuitry with its battery in the bell box cannot be tested other than by either open-circuiting or short-circuiting the bell wire. The bell is thus self-actuated and caused to ring, but it then cannot be stopped from the control box. Such a test is obviously impractical, and this inability to make routine tests is its weak point. Although self-actuating bell circuits are specified by BS 4737, it seems preferable to use an ordinary bell and take extra care over the wiring by protecting and concealing it over its whole length.

System switching

The traditional method of switching the system is by means of a key-switch on the control box. Some manufacturers make 2000 different key profiles, so the chances of the same one turning up in the same area are remote. Most key-switches for single-zone units have three positions, 'off', 'test' and 'on'; for twin-zone boxes the positions are usually 'zone 1', 'off' and 'zones 1 and 2'. With the first one the switch must be taken through the test position to reach 'on', while with the second it is turned once either to the left or to the right to select the required coverage. With multi-zone units, separate switches, which may be toggle or press buttons, select the zones to be activated. The main key-switch may then be inscribed 'part on', 'off' and 'full on', so that a pre-selected group of zones may be conveniently switched in the 'part on' position.

Keys are removable in any position. Most control box makers supply only two, and permit no reordering of duplicates in order to maintain security. If more are required they can usually be cut locally from one of the originals, although this does compromise security.

Most people find keys a nuisance, especially when so many different ones have to be carried. With this in mind, many control boxes are operated by a key-pad similar to the buttons on a calculator. There are now about three key-pad models for every two key-switched ones. A three- or four-digit code is entered to set and switch the system off. A code may be already built into the unit, but in most cases it can be changed by the user to one of his own choice. Some models have both a key-pad and a key-switch so that both or either can be used if required. A few use six-digit codes for extra security.

A three-digit code can offer 1000 possible numbers including 000, whereas a four-digit one offers 10,000. The latter is thus ten times more secure as it is ten times less likely for an intruder to hit on the correct code by chance. With some panels a digit cannot be used more than once in a particular code, so the possible numbers are reduced to 720 for three-digit codes and 5040 for four-digit codes.

If an intruder manages to gain entry without setting off the alarm he may attempt to switch it off by trying different codes on the key-pad. However, if the wrong code is entered with some panels, the entrance timer starts, if it is not already running. If the occupier has miskeyed the code, this gives him sufficient time to get it right before the alarm sounds. Yet it would be insufficient for an intruder to try a list of possible numbers.

However, he may have time to try a few possibles. So in choosing a code, care must be taken to avoid the obvious. Many users choose birth dates or anniversary dates, but these could be easily discovered by someone planning a burglary. Consecutive numbers forwards or backwards should also be avoided although they may be easy to remember, as any such sequence could be covered by keying 0–9 and back again. A row of the same digit, especially 999, is another one to steer clear of.

There is no harm in choosing a number that is easy for *you* to remember because it has some other significance, but make sure it is one that no one else knows or could guess.

With some models the system goes into the alarm state if more than a certain number (13 in one model) of digits are keyed. This permits a second or third try after a miskeyed entry of a four-digit code, but does not allow any more. The intruder would thus have to guess it correctly by the third try, a very unlikely chance.

Industrial key codes

Some industrial models have key-pads that respond to more than one code. One could be a cleaner's code that will set the system but not switch it off. Another might be an engineer's code that enables routine testing to be carried out without divulging the main code. Thus if any of these codes should fall into the wrong hands, security is not compromised.

A few models have a number of entry codes each of which can be assigned to different employees. Should one leave the company, his code can then be deleted or changed without everyone else's being affected. With these a master code, when entered, allows any one of the others to be altered, so enabling a change to be made without delay. Another variation is to allow the system to be set by entering only the first two digits of the code. This can be more convenient when leaving the premises, and permits unauthorized persons to set the system without knowing the full entry code.

Some panels have a *duress code* in which the last digit is different from the normal one. If entered under duress, it initiates a personal attack alarm. This may sound the full alarm, or it may signal to a distant point, so giving the impression that the alarm system has been switched off. The hostage thereby avoids retaliatory violence, while alerting the authorities of the situation.

Though making such facilities possible, many key-pads do have one disadvantage. These are the type that does not have actual press keys, but has a touch-sensitive membrane on which the numbers are inscribed. They should be operated with the front of the finger but as they are mounted vertically, most users poke them with their fingertips. They thus become indented by fingernails, especially women's nails which are usually longer.

This indentation can betray the numbers of the code, though not their sequence. There are but 40 possible combinations of four digits and only 24 if the same digits cannot be repeated, so the sequence would not take long to discover if the panel has no excess number limit. Another way of finding the four digits of the code is to dust the keys with chalk powder, whereupon it will stick to those having a greasy deposit from frequent touching. These ploys are of little help though where several different codes are employed, as most numbers then show signs of use.

Some control boxes permit the connection of remote control consoles. These are small units that can be stationed at key positions within the protected area and allow all the functions of the master controls to be carried out. With one model, up to four remote units can be installed. None of the features in this section are applicable to domestic systems.

Indicators

Some control units look formidable, with a row of lights to indicate various things, along with the control switches, a key-pad and sometimes a read-out screen. They are not quite so bad as they look. First there is the ubiquitous power lamp which is on all the time the power is connected. Apart from that, most of the others are associated with one of the circuits: anti-tamper, exit, personal attack, and one each for the zones. These come on when there is a fault in that particular circuit at switch-on or test.

In the event of an alarm, the light shows which of the circuits has been triggered. In some cases they all come on momentarily at switch-on to indicate all are functioning, then go off. Some units have a 'clear' light which comes on when testing to show that all is well.

Some microprocessor-controlled panels have a read-out screen which gives the status of the various detection circuits, prompts the user to take certain actions, and provides a record of the last alarm events with day and time. Some record up to 500 alarms, but if there were ever that number it would be better to move!

Most domestic control boxes have a limited number of indicators as they are usually single-zone units.

Output facilities

These vary between models. Most control units supply a maximum of 1 ampere (A) to the sounder circuit, but some provide only 0.5 A (500 mA). It is recommended that normally the load be no more the 80 per cent of the maximum rating. As bells can take up to 300 mA and sirens from 0.5 A to 3 A, it is important to ensure that the control unit will supply sufficient current, especially where two or more sounders are to be used. In such cases the current rating of each sounder is added together to give the total, which must be less than the maximum specified for the control box.

All modern control units now have output timers that cut off the bell after a pre-selected period. The recommended time is 20 minutes. This has been introduced in response to widespead complaints by neighbours about bells sounding for long periods when the owner has been absent. This is particularly annoying when triggered by a false alarm as most of them are. Other internal sounders that cause less annoyance may continue until switched off, as well as the external flashing light.

Many control units have a further facility in that they contain or will trigger a communications device that dials 999 and give a recorded message, or one that has a direct line to the communication centre of a security firm or British Telecom.

Some of these models enable the sounders on site to be delayed to allow the police to catch the intruder. The wisdom of this though is dubious, as much damage could be done and the culprits depart with items of value before the police arrive. Really, it is best to keep them out, and the most practical way of doing that is to startle them and scare them off.

Power supplies

The power for running the system is taken from the mains supply, with back-up batteries which switch in automatically if the supply fails. The control box must be permanently wired to a suitable mains connection box. It should never be supplied from a plug and socket, as the socket could be switched off or the plug be pulled out. This could be done innocently by someone using say a vacuum cleaner, just as it could deliberately by an intruder.

The power supply is probably the weakest point in the whole system. A breakdown in the public supply, likely in rural areas in bad weather, or a circuit-breaker being tripped by a fault on the same house circuit, leaves the system at the mercy of the standby batteries.

These have a limited capacity, of the order of hours rather than days. BS 4737 stipulates a minimum capacity of 8 hours, which is woefully inadequate and is insufficient protection for even a weekend away, quite apart from longer holidays when many burglars are most active. Most standby batteries exceed this capacity, but the time is still limited, being $2\frac{1}{2}$ days on average for a typical control panel and rechargeable battery. Even this is reduced if infrared or other space protectors are used.

Industrial models are the biggest problem because of the current required by the various control circuits. The anti-tamper loop takes current as well as the detection loop, and furthermore is active for 24 hours. Multi-zoning, which is not usually found with domestic systems, multiplies the loop current and requires current-consuming indicators. While some domestic systems also have limited standby battery life, others have such a low operating current that dry batteries will run the system for a year or more. The Sureguard design described in Chapter 14 will run for up to two years on dry batteries.

The length of time that a system will continue to function after the mains supply has failed will thus depend on the control unit, the number and type of sensors, and the rated capacity of the battery. Strangely, it is a factor that is rarely mentioned in system specifications. It is assumed, it seems, that mains supplies will never fail, or if they do will be restored quickly – an assumption that could prove dangerous!

Reset

Most units either have a reset button to clear the alarm circuits after triggering, or reset automatically on switching off. For those connected to a 999 dialler or other outside communicator, it is the normal practice for the user to be unable to reset the system, which can only be done by an engineer. The reason is to prevent false alarms. If the original alarm was false owing to some malfunction, merely resetting and switching on again without finding the trouble will only produce another false alarm and a cool reception from the local constabulary. When an engineer is called out after a false alarm, he makes a thorough check to find and clear the cause before resetting the system. The chance of a further false alarm is thereby greatly reduced.

When the bell timer has expired after an alarm, some systems reset but inhibit those detection circuits that are still in an alarm state; this stops them re-sounding the alarm, yet the rest of the premises continues to be protected. The facility is optional in many models and can be suppressed if desired.

There are now a large number of manufacturers each offering a range of units in a competitive market. This breeds innovative ideas and new or different solutions to old problems; it can also breed gimmicks. So it may be expected that units will appear offering something different, and this is good for progress.

However, be wary of rushing into something new just because it is new, or seems or is claimed to be better than anything ever produced before. Maybe it is, but also there may be serious snags that will only be discovered later and in certain situations. Many a manufacturer has launched a seemingly promising product only to find it a disaster and themselves bankrupt. Customers who bought it were left high and dry with no spares, maintenance or advice. Where security is involved it pays to be cautious.

5 Sensors

Sensors are a vital part of any alarm system. These are the guardians that respond to any disturbance caused by an intruder and trigger the main alarm circuit. Having done this they play no further part in the operation, as the latching circuit in the control unit takes over to keep the alarm sounding. Hence any attack on the sensor or its wiring after the alarm has sounded is of no avail in silencing the system.

The sensors must be perfectly reliable and operate every time they are actuated. However, they must also operate *only* when triggered by an intrusion and not generate false alarms as a result of wind, traffic, vibrations and other causes.

There are two broad classes of sensors. The first consists essentially of switches of various types. They are usually fitted to defend the perimeter of the premises, such as windows and doors, although some such as pressure mats are used within the protected area.

The second class protects space by detecting movement within the area. These sensors really back up the perimeter defences. In this chapter we will consider the first class, and in the next the second. Not all those described would be used for most domestic systems, but they are included as they may help with special situations.

Switch contacts

First a few facts about switches. Either the contacts are *normally open* (NO) and are closed when they are actuated, or they are *normally closed* (NC) and are opened when actuated. Some sensors have two pairs of contacts; one pair opens while the other closes. Others have three contacts, one of which is common and switches between the other two. These are called *changeover* (CO) or *single-pole double-throw* (SPDT) switches. There are also double-pole double-throw switches (DPDT) which have six contacts and are really two SPDT switches operated by the same actuator. These though are not often encountered in security systems.

Normally-closed contacts are used in continuous loop arrangements where a current constantly circulates when the system is armed. Actuation opens the contacts and stops the current which then triggers the alarm. As we have already seen, this is the main method of detection, so chosen because cutting or breaking the wires also stops the current and

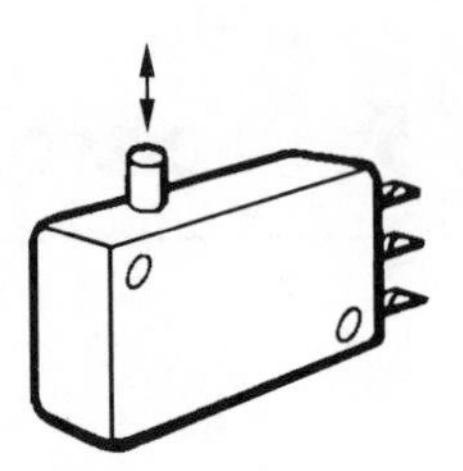

Figure 14 *Basic microswitch. Arrows show direction of operation.*

sounds the alarm. There is thus a built-in security safeguard against damage or deliberate wire cutting.

This obviously cannot give warning of interference when the system is switched off. A second loop is therefore run to all sensors in commercial systems which carries current 24 hours a day, even when the system is not armed. Any tampering with the wiring thereby gives an alarm indication at any time it occurs. This loop is not switched in any way by the sensors; it merely runs to terminals on them which serve as connectors. Most sensors thus have at least four terminals, two to the switch, and two internally connected together for the anti-tamper loop. These are ignored in home systems that do not have an anti-tamper loop.

Normally-open sensors trip the alarm when they are closed. These are mostly pressure mats which are difficult to make in a normally-closed mode. However, an anti-tamper loop is also run to these in commercial systems, so that they are protected against daytime interference.

The changeover type of sensor switch offers even higher security against tampering by using both normally-open and normally-closed operation for the same sensor. Actuation of either mode triggers the alarm, as does tampering with the anti-tamper wiring. These are not usually considered necessary except for premises where very high security is required owing to the possible attentions of determined and knowledgeable professional burglars.

Microswitches

The simplest type of switch is the microswitch (see Figure 14). As its name implies, it is a small switch which can easily be fitted in door or window frames, display cases, safes, cupboards, cabinets and drawers and under desk lids. It is commonly used in bell boxes and control units in conjunction with the anti-tamper circuit which is actuated if the cover is removed.

The basic microswitch is actuated by a plunger which is sprung so that it is depressed to operate and returns to its rest position when the pressure

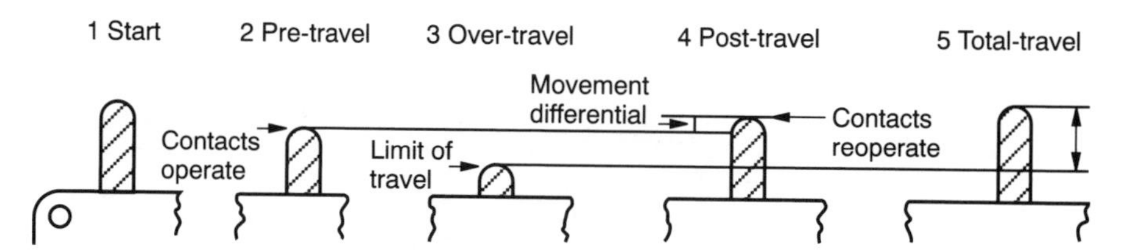

Figure 15 *Microswitch operation chart showing operating points.*

is removed, just like a bell push. The amount of travel required by the plunger and the pressure needed vary considerably from one switch to another, and selection is determined by the application.

Some makers give full mechanical specifications as to the various parameters of the plunger travel (Figure 15). These are necessary for the selection of a switch for a particular purpose, especially when the movement involved is small. From the start position, the amount of plunger travel before the switch contacts operate is known as the *pre-travel.* Movement from this point onward is termed *over-travel,* and while a certain amount is necessary to ensure that the contacts have in fact operated, it should not exceed the stipulated *limit of travel.* On release of the plunger, the travel from the limit to the point where the contacts are re-operated is designated the *post-travel.* It may be noted that the re-operation point is not exactly the same as the initial operating point, it being nearer to the rest position: the difference between the two is known as the *movement differential.* The distance between the at-rest and limit-of-travel positions is logically described as the *total travel.* All this may sound unnecessarily involved for a simple press switch, but it is essential to observe these limits if the switch is to function as intended and maintain reliable operation over a long life.

The forward stroke is of little interest for alarm applications as the sensor is normally held in the depressed position, although it must not be exceeded. It is the return stroke which requires consideration. Post-travel must not be too small, otherwise vibration and other disturbances may actuate it. Even a door with a loose catch may initiate a false alarm as it may not always be shut closely.

The operating movement of some microswitches is of the order of 0.3 mm which is obviously far too critical for normal door or window operation. These types are designed for much more sensitive applications. Long travel can be selected for trophy case doors or desk lids as these need to be opened wide to obtain access. This will reduce the possibility of false alarms. Shorter travel can be selected for more precise and critical operation such as safe doors.

An external attachment often supplements the basic microswitch to

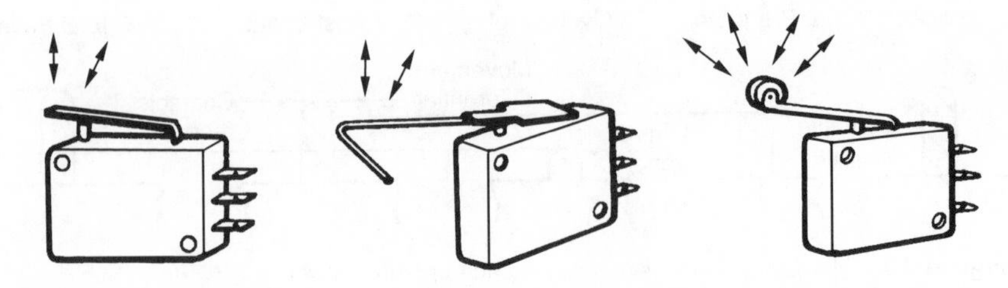

Figure 16 *Microswitch attachments: left, leaf or lever actuator; centre, wire actuator; right, roller actuator. Arrows show operating directions.*

alter the mode of operation (Figure 16). There are three main types: the leaf or lever attachment, the wire actuator, and the roller attachment.

The leaf or lever attachment is hinged at one end and passes over the plunger so that movement at the free end actuates it. The effect is to amplify the movement needed to operate the switch, and the longer the lever the greater the movement required. Various lengths from 18 mm to 76 mm can be obtained.

Operation of the basic microswitch must only be by vertical pressure on the plunger; any movement with a sideways or wiping action could deform, jam or break it. However, such movement is permissible with a lever, but it must be in one direction only, and that is away from the hinge for the downward stroke. Some makers have models with adjustable hinge positions so that leverage and travel can be varied accordingly.

The second attachment is also a lever but instead of a flat blade it consists of a stiff wire with a right-angled bend at the free end. The operation is the same as for the bladed lever but the wire serves more as a feeler. A thin twine trip cord could be attached to the wire or it can be used wherever a sensitive touch is required.

Neither of these can be used where a sliding contact is needed. For this, the roller type of attachment is necessary. Rollers are usually mounted at the end of levers, but they are also available fitted into the actual plunger. They have a particular application for sliding doors and drawers. The advantage is that they can fitted an inch or so from the end of the frame so that they will not be actuated until the door or drawer end reaches that point.

If then the door is not fully slid home as often happens with sliding doors and drawers, it will not trigger a false alarm. Yet if opened enough to gain access, the switch will be actuated. Thus both high security and immunity from false alarms can be achieved. It should be noted that while for many applications the magnetic switch to be described later is preferable to the microswitch, sliding surfaces are best served by roller microswitches.

Most microswitches have a pressure rating ranging between 0.25 and 16 oz. Two figures are given, an operating pressure and a maintaining pressure. For most applications the pressure is not too critical, but for some delicate operations the required pressure would need to be low. If too great it could actually hold off the actuating surface and prevent the switch operating.

Various fixing arrangements are possible as there are different fixing hole placements. Some have holes passing through the sides for sideways fitting, other have holes through the top plate for flush mounting, while others have a screwed bush through which the plunger passes for single-hold panel mounting.

In spite of their size, microswitches have high current ratings, but this is usually of little importance in latching alarm circuits as these pass very small currents. This could be useful, though, where the switch is required to operate an alarm bell or other load device directly. Magnetic switches cannot be used because of their low current ratings, so the microswitch is the solution. Most are rated at 5 A which is more than adequate for most purposes, but some go up to 20 A.

Operating life is normally high and is usually specified by the maker. The lowest is some 100,000 operations, but typically it extends up to 10 million. These sorts of figures are of interest to designers of industrial counting or sensing equipment but they would never be approached with normal alarm applications and so they can be ignored.

The majority of microswitches are made with SPDT contacts which enable them to be connected in either the normally-open or the normally-closed mode. They can also be incorporated in a security door lock to remote-switch an alarm system as this often requires SPDT contacts. An alternative that is sometimes used is to house the microswitch in the door frame cavity so that it is actuated by the lock bolt. This avoids the need to run wiring to the door itself which requires a special security loop connection across its hinge side. The disadvantage is that the microswitch needs careful adjustment so that is is fully actuated by the bolt, but is not driven beyond the travel limit.

Magnetic switches

A disadvantage of microswitches is that they are vulnerable to physical damage. The plunger or lever attachment can suffer deformation and jamming or even breakage by physical impact. Inaccurate positioning can cause the plunger to be driven beyond its limit of travel and so result in damage. If exposed in a door frame or a similar position, it could be

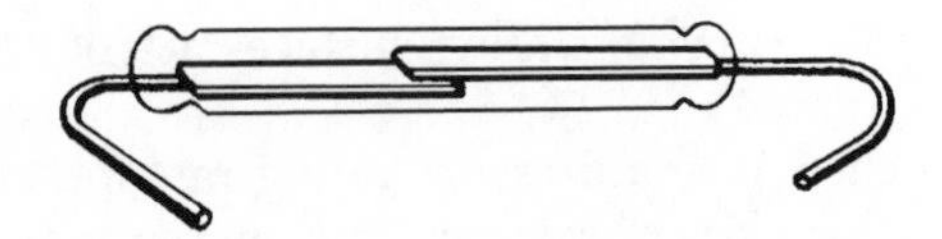

Figure 17 *Magnetic reed switch. Reed contacts are sealed in a glass tube and are actuated by an external magnetic field.*

tampered with by sticking the plunger down with chewing gum or superglue. This could not be revealed by a normal test as the switch contacts would be in their normal position. The damage would only come to light at the next walk test, which may be too late. All these are possibilities unless the switch is well protected by being mounted invisibly or in a non-accessible position such as inside a desk, cabinet, or lock.

The problem arises because mechanical movement is required to actuate the switch and the actuating portion is open and exposed. Ideally, a sensor should have no external moving parts and should be completely protected by its surroundings. These characteristics are closely achieved in the magnetic reed switch. It consists of a pair of leaf contacts completely sealed inside a glass tube. Each contact is supported at opposite ends of the tube, and they overlap at the centre where the contact is made (Figure 17). When the device is located within an external magnetic field the contacts become temporarily magnetized and attracted to each other, and so they close. When the field is removed, they are demagnetized and spring apart.

The tube is encapsulated inside a plastic case which is mounted in a door or window frame. A matching case containing a bar magnet is fitted to the door or window so that when it is closed the magnet lies adjacent to the switch. The contacts are thus magnetized and close, so they can be connected in series with a closed loop. When the door or window opens the magnet moves away and the contacts open.

We should here clear up what may seem a confusing description. The device is described as normally closed, because when it is installed and the system is on guard with the door shut, the contacts are closed. This of course is because they lie in the magnetic field of the mating piece. However, when the switch is free before it is installed, the contacts are normally open, because they are not influenced by any magnetic field. Care must be taken over this when installing, because the switch terminals will show open-circuit if checked with a continuity tester, whereas the linked anti-tamper contacts show a short-circuit. It is easy to be confused by this and connect up the wrong pair.

The magnetic reed switch offers numerous advantages over the microswitch for protecting doors and windows. Having no external moving parts it is less vulnerable to damage and is more difficult to defeat. It does not rely on actual contact with the door or window and so is not affected by vibration or minor irregularities such a wood swelling or shrinking. There are no restrictions as to the angle of travel of the magnet either toward or away from the switch. The only possible disadvantage is that two units must be installed instead of one.

Positioning and distance between the units is not as critical as with the microswitch and its actuating surface. In fact there can be quite a gap between the door and its frame without causing problems, a distinct advantage with many properties.

The distance at which the magnet will influence the switch varies according to the type of switch and strength of the magnet. Two distances are quoted by the makers: *operating distance* and *release distance*. The former is the distance at which the approaching magnet will cause the contacts to close, and the latter the distance at which the receding magnet will allow them to open. The release distance is roughly one and a half times the operating distance, but in some cases can be twice as far. For door and window sensors it is the release distance which is the significant parameter.

Typical release distances can be from 10 mm to 35 mm. These represent the distances that the door or window will open before the alarm is triggered, assuming that the units were in close contact with each other to start with. If there was a gap, which is virtually certain, the gap would have to be subtracted from the release distance to find the opening distance at which the alarm would sound. The gap must always be less than the operating distance otherwise the contacts would never close.

Distances quoted by the makers are subject to a certain tolerance between individual units and are measured with a new magnet. As magnetism is lost with age, distances will be reduced as time passes. To ensure reliable operation at all times and with all units of a particular type, minimum distances should be reduced by 25 per cent.

Switch contacts are of precious metal and, being sealed in, are unaffected by atmospheric moisture or pollution. They have therefore an extremely long life, over 100 million operations, which is ten times that of the average microswitch. So if a door was opened and shut once every 10 seconds for 8 hours a day, 7 days a week, the contacts would last for over 95 years. Although such a life expectancy is never likely to be reached, it gives some idea of the inherent reliability of the magnetic reed switch, and another reason why it is so eminently suitable as an alarm system sensor.

As with the microswitch, some reed switches are available with SPDT changeover contacts so they can be used in the normally open as well as or instead of the normally closed mode. These are not often used except for systems requiring the highest possible security.

Figure 18 *Circular magnetic switch and matching magnet.*

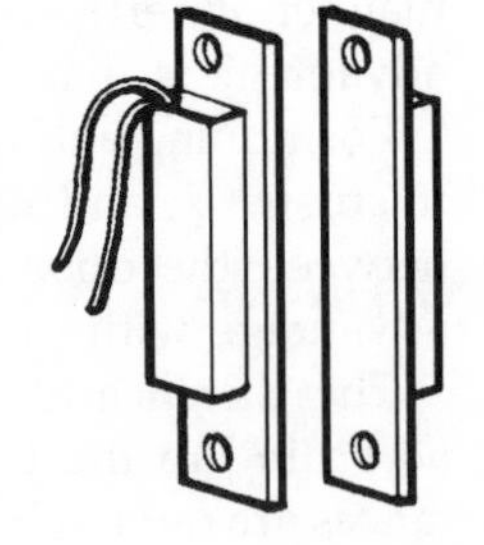

Figure 19 *Flush-mounted shallow rectangular magnetic switch and matching magnet.*

The current rating is limited and the magnetic switch should not be used to directly connect a bell or other signalling device. If such a circuit is required, a microswitch should be employed.

Encapsulation

There are several different types of encapsulation to suit various mounting requirements. One is the circular flush-fitting type, which is inserted into a hole drilled in the door frame, a flat disc forming a head which lies flush with the surface. The magnet is fitted into a matching unit that is similarly sunk into the edge of the door (Figure 18).

The result is inconspicuous and the only woodwork needed is the drilling of the holes to accommodate the units, with another smaller hole intersecting with the bottom of the switch hole to take the wiring. As the units are about 35 mm long, there must be at least this depth of wood to receive them. This may not be the case with shallow frames or with windows or glazed doors. Another factor is that as the mating surface area is small, the switch and magnet must be accurately aligned with very little tolerance for misalignment.

Another type for flush fitting is the shallow oblong variety (Figure 19). As these are only a few millimetres deep they can be installed where the wood is not very thick such as in window frames. A cavity must be chiselled out to accommodate them and they are supported by screws through the overlapping faceplate. The faceplate is normally raised above the wood surface, but if a really unobtrusive job is desired this too can be recessed into the woodwork. If puttied in and painted over it can be indistinguishable from the surrounding surface. As the mating area is

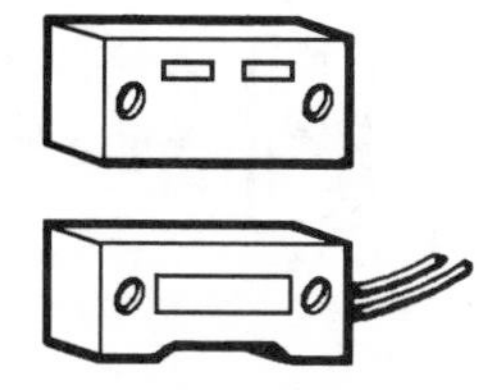

Figure 20 *Surface-mounted magnetic switch and magnet.*

quite large, accuracy in lining up the switch and the magnet is not so important.

There is also an oblong surface-fitting type as shown in Figure 20. These are mounted on the surface of the frame on the opening side, and the magnet is set on the door surface so that they meet edge-to-edge. Being visible and accessible these are much less secure than the flush-mounting varieties for commercial premises where tampering is a possibility. They are sometimes supplied with DIY home kits to make fitting easier by saving the drilling involved with the flush-mounting types. As they are mounted on the inward side of the door which opens into the protected area, they cannot be tampered with from the outside, so security is not greatly sacrificed. Appearance is thus the main factor, as the flush-mounting units make a neater job.

With metal windows there may be no alternative, as it is virtually impossible to fit a flush-mounting type with these. However, with some double-glazed aluminium windows there is sufficient room in the hollow section from which they are constructed to accommodate a magnet and a switch in the frame.

Although unusual, it is not impossible to defeat a reed switch by using an external magnet. First of all the intruder needs to know its position in the door frame. This can easily be discovered by running a compass around the frame. The magnetic field from the associated magnet will turn the compass needle and give an accurate indication of the switch location. Now he needs to use a strong external magnet to immerse the reed switch in its magnetic field. This can be fixed temporarily to the door frame in some way, and then the door can be forced open. The magnet thus holds the reed switch closed and no alarm is triggered.

This method is unlikely to be used by the average home intruder because it requires knowledge and skill instead of brawn, but it could well be employed by the professional thief. If the premises contain valuables likely to be attractive to the professional, this possibility cannot be discounted and good back-up protection should be provided.

One method of increasing security against this trick is to fit two switches to each perimeter door frame, one high and the other very low. On discovering one, possibly the high one, the intruder would be likely to assume it was the only one and not search further, especially right down

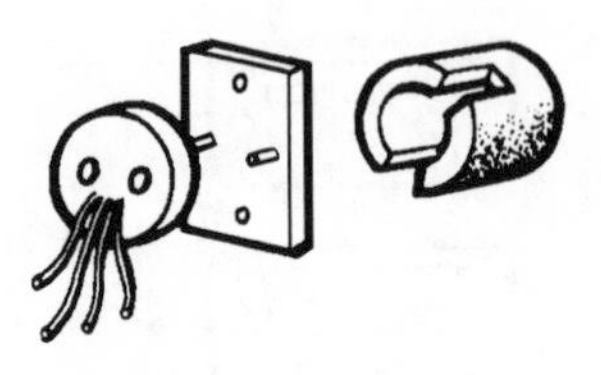

Figure 21 *High-security double-reed switch. Can only be actuated by both poles of a magnet, hence the bifurcated magnet shape.*

to floor level. He would thus be caught by the one he missed. If he did find the two, the chances are he would only have one magnet and so would be thwarted anyway.

Another way to reduce the possibility of defeating the magnetic switch by use of an external magnet is to use a high-security type which contains two reed contacts that need energizing by both poles of a magnet (Figure 21). The mating magnet is cylindrical but is bifurcated by a central slot with opposite poles being formed in each section. These sections must be lined up to the two sets of reeds in the switch when installing in order to actuate them. The use of an external magnet or magnetized strip will not operate the switch and so it cannot be neutralized by this means. This type of sensor is recommended for installations where the highest security is required.

It should be noted that with all magnetic switches the close proximity of ferrous metal such as metal window frames may affect the operation, particularly the operating distances.

Connections are of two types, those having terminals and those with lead-out wires. The terminal type is the easier to fit and is quite satisfactory providing the terminals are well tightened. The lead-out wires of the other type must be soldered to the circuit wiring; although this type is less convenient to install, the soldering ensures permanent joints that will not work loose and cause baffling intermittent false alarms. Some are obtainable with 3 ft (1 m) lead-out wires and are designed for aluminium or UPVC windows, being housed in either aluminium or white plastic casings. The long wires enable a joint to be made unobtrusively away from the window frame.

Some applications need a particularly robust unit that will stand up to adverse conditions. Roller shutters, up-and-over garage doors, and bank shutters call for specially strong encapsulation. Switches are available sealed in aluminium cases that are flat on one side but pod-shaped on the other (Figure 22). The depth is a mere 0.5 in (13 mm), and the unit is designed to be screwed to the floor, using plugs if the floor is concrete.

The switch can be driven over by a car without damage, and there is one model which it is claimed will not suffer if driven over by a fork-lift truck! The magnet is housed in an aluminium casing that is fixed to the door, and is angled inside the case to minimize the effect of a steel door.

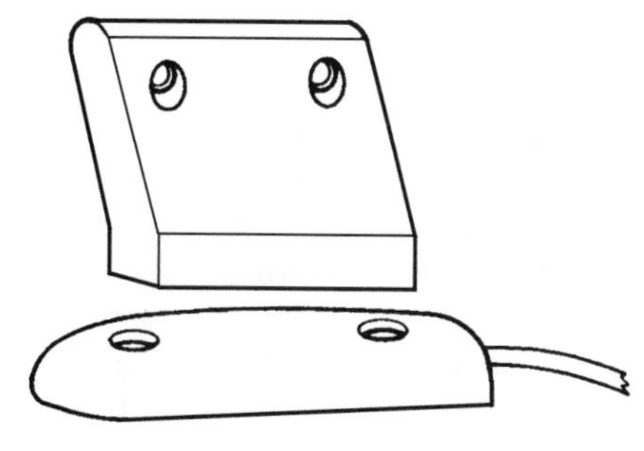

Figure 22 *Garage door switch with magnetic actuator.*

Resistor-bridged contacts

The dual-purpose loop, you may remember, uses a resistor across each pair of contacts. An end-of-line resistor terminates the loop. When the system is in the 'day' mode, a current circulates around the loop, monitoring that it is still intact. If contacts open when doors are opened, the resistor maintains the continuity. When the system is switched to the night mode, all contacts are closed and the end-of-line resistor is the only resistance in the circuit. The control panel senses its value, so if a contact is opened and the resistance changes, an alarm is triggered (see Figure 13). This dispenses with the separate anti-tamper loop and also permits normally open sensors to be connected across the same loop, so saving extra wiring for these.

Accommodating the resistors across each sensor could be inconvenient especially with flush-mounting units, so special ones are available with a resistor built in. These should not however be used with conventional closed-circuit loops as current still flows in the loop when the contacts are open. With these, the control unit needs a complete cessation of current to trigger an alarm.

Radio sensors

Another system is the radio-controlled alarm system consisting of sensors and control unit. The sensors are conventional switches which can be surface mounted to doors and windows. However, instead of being wired to the control unit, each sensor has a tiny radio transmitter powered by its own internal battery. When the sensor is operated a radio signal is transmitted that is picked up by the control unit which then activates and latches an alarm in the usual way. A feature of the system is that, having no wiring, it is portable, and so can be quickly installed in temporary premises and just as quickly removed. The disadvantage is that there is no anti-tamper system and only limited monitoring is possible owing to the small internal battery capacity.

Transmitters are available that can be fitted to existing space protection devices, and receivers can be fitted to ordinary control panels. Thus a system can be part radio (termed free-wired) and part conventionally wired (called hard-wired). Sensors that pose difficult wiring problems such as to outhouses or to odd remote windows or doors could be of the radio type and so eliminate the wiring, while the rest of the system was hard-wired. The range is up to 200 ft (60 m), but this can be extended by an aerial.

A further useful feature of the radio system is the provision of radio panic buttons. These can be used anywhere within range, which is the same as for the sensors, and they can be carried inside a pocket or handbag. Thus protection is afforded while working in and around the house. While a radio system would not be used in the normal domestic installation, it could be very useful in some circumstances.

Pressure mats

We have referred to these several times already, but now we describe what they are and what they do. They are rectangular flexible mats that come in a variety of sizes but usually with an area of 2–6 ft^2 (0.2–0.6 m^2). The usual method of construction is a sandwich of two sheets of metal foil separated by a layer of perforated foam plastic. When pressure is applied to the mat the foil sheets make contact through the foam perforations, and when it is released the foam separates them again. A plastic covering seals the mat against dirt or moisture. Care must be taken when installing not to puncture this cover as, although the hole may have no immediate effect, it could cause trouble later.

There are four lead-out wires which must be soldered to the circuit wiring. Two go to the foil sheets and are connected to the normally open detection circuit, while the others are straight-through links which connect to the anti-tamper circuit. These can be ignored and cut off in a domestic system having no anti-tamper circuit. To identify them, the switch wires are usually stripped for a few millimetres, while the anti-tamper wires are not.

The thickness is typically about $\frac{1}{4}$ in (6 mm), and the mat should be installed under a carpet and underlay. A slight bulge may result, but this will soon bed down and become almost imperceptible after a while. In domestic systems the pressure mats would be placed on a stair, in a hallway, and in front of video recorders, TV sets, hi-fi, and bedroom dressing tables where jewellery may be expected. In short, they can be positioned anywhere an intruder is likely to step.

The pressure mat can serve areas where perimeter protection is difficult or of a low order of security, such as a bay window or a patio where the windows are too numerous or difficult to individually wire. Really, though, they should be regarded as a second line of defence. If intruders do manage an entrance without actuating a perimeter detector, they may step on a pressure mat. But these should not be relied on: the principal object of the system is to protect the perimeter and keep intruders out. There are other sensors for protecting the interior, as we shall see in the next chapter, but mats have the advantages of being cheap, being less prone to false alarms, and consuming no current.

The amount of pressure needed to actuate a mat is rarely quoted by the makers. It is usually about 2.5–3.0 lb/in^2 (0.18–0.21 kg/cm^2). The average total area of a man's shoe sole and heel is about 25 in^2 (160 cm^2), which for a 10 stone (64 kg) weight gives a pressure of about 5.6 lb/in^2 (0.4 kg/cm^2). A woman is on average lighter than a man, but the shoe area is much less, so the pressure is actually greater. A child's weight too is less, but so is its shoe area. A large dog, being heavy and having a small paw area compared to a human foot, could actuate a mat, but there is little risk from a small or average-sized animal.

These figures assume that the foot is laid gently and evenly on the mat, but in walking this is not the case: the heel comes down first with some momentum, and so the force acting on the mat is much greater. There is little fear then that it will not respond when trodden on. The weight of a cat is insufficient to actuate it unless the animal jumps down on to the mat, so the resident mouser is unlikely to be the cause of false alarms. This is a further advantage over space protection devices which can be triggered by pets.

One thing that must be watched, though, is that movable items such as chairs and tables are not inadvertently left standing on the mat. After a mat has been installed for some while the tendency is to forget it, and so there is a strong possibility of something being moved on to it and left there. The weight of the offending article may be insufficient to immediately create a contact, but it could settle after a few hours and trigger the alarm – in the middle of the night! This has been the cause of many a false alarm.

A drawback with pressure mats is that, being normally open devices, they cannot be part of a closed loop and so cannot be tested along with the other sensors each time the system is switched on. They can only be checked by a walk test, that is by actually operating the device with the control unit in the armed or test mode.

Use only the best quality pressure mats. Cheap ones have inferior foam insulation between the foil sheets which breaks up after a while, especially if located in a well-trodden position. If the carpet is thin or there is no underlay, the pressure mat will not last long. Even if they are well protected, be prepared to replace them from time to time.

Window strip

Various methods can be used for protecting window glass. One of the most common is strips of metal foil fixed across the inside surface of the glass. These are usually made of aluminium but lead is available for use in areas where high atmospheric pollution could corrode the aluminium. The strips are supplied in rolls in various widths, $\frac{1}{8}$, $\frac{1}{4}$ and $\frac{3}{8}$ in (3, 6 and 9 mm) being the most common. They can be either self-adhesive or non-adhesive, but the self-adhesive type is the most convenient to apply.

The foil is run along a vulnerable area of glass, and is connected into the closed-loop circuit, the 24-hour panic button circuit, or an anti-tamper loop. If the glass is broken, the foil is severed and the loop is open-circuited. If connected into the anti-tamper or panic button circuit, protection is afforded for 24 hours a day. Unlike other sensors it causes no inconvenience or likelihood of false alarms when active unless someone accidentally breaks a window!

It is of special value for shop premises to guard display windows, but it also has uses in the home. In particular, front doors that have glass panels are vulnerable, and can be so protected.

Special self-adhesive terminal blocks are fixed to the glass at the side of the pane. These make contact with the foil strip and have terminal screws for connecting the wiring. Making-off strip is available for bridging two panes across a window frame.

The normal method is to run two horizontal strips at roughly one-quarter and three-quarter positions up the glass. This leaves the centre clear but gives adequate protection and can even be an embellishment. By using two strips in this way the circuit wires can be connected at the same side of the window, the free ends being linked on the opposite side by a wire up the frame.

One feature of metal foil on glass is that it can be seen from the outside. An intending intruder is thus informed that the glass is protected by an alarm and is deterred from attempting a break-in through it. This is far preferable to the alarm being triggered by some invisible detector after a breakage, as it saves the not inconsiderable cost and inconvenience of replacing the glass. Deterrence is always the best form of defence.

There are obvious problems associated with fitting metal foil to windows that open. The connecting block would have to be fitted to the window on the hinge side and flexible leads bridged over to another block on the frame to connect the circuit wiring. It is better then to use a magnetic sensor for all opening windows.

While in most cases metal foil offers good protection to fixed glass, it is not unassailable. As noted before, a determined intruder could use a glass cutter to cut out a section avoiding the metal strips. He would still have to make an entry between the remaining glass strips in the frame,

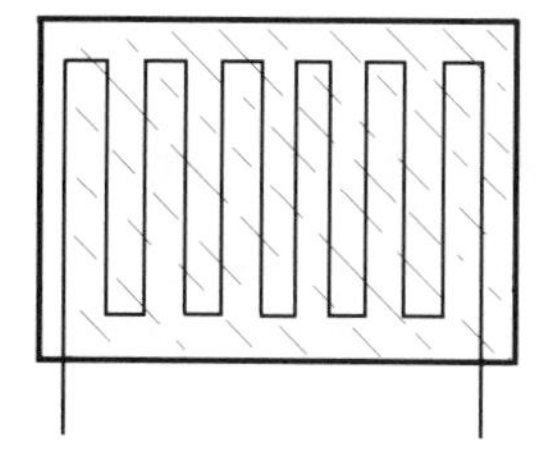

Figure 23 *Wired glass consisting of two bonded sheets with silvered wire at 2 in (50 mm) spacing.*

but if the window and hence the gap is large, and he taped over the cut edges, he could get in with little risk of injury.

One remedy would be to run the foil strips closer together or put further strips across the glass, but this could reduce visibility to an unacceptable degree. An alternative is to use wired glass as shown in Figure 23. This is a special glass consisting of two sheets sealed together with a series of silver wires across almost the complete width. The wire is very fine and is hardly visible, similar to a car rear-window heater, but being spaced about 2 in (50 mm) apart gives a high degree of protection. As with the foil, the wire is connected to the closed or 24-hour loop. This is obviously more expensive than foil which can be fitted to existing glass, and is hardly necessary for most domestic applications, but it affords high security where that is needed, with good visibility.

A cheaper alternative would be to put a pressure mat underneath the window which could not be avoided by anyone climbing through.

Roofs and ceilings

Roofs and ceilings are often overlooked as a means of entry but, especially with bungalows and single-storey extensions, they can afford an easy way in for an intruder. Tiles can be quickly lifted off and roofing felt torn open, or with flat roofs the bituminous felt covering can be soon penetrated. Next comes the heat insulation, which is no problem to remove, and then the ceiling, which is easily broken through.

With bungalows having the usual pitched roof, there is no need to penetrate the ceiling. Most have a trapdoor leading to the roof cavity which may be just lifted out, or if hinged and secured on the underside by a catch or bolt can usually be broken open by a hefty kick.

The exit poses little difficulty as there is usually a door or window that can be opened from the inside. Even if this sets off the alarm, the thief is on his way out and is well away by the time anyone is alerted.

So, extension ceilings can and should be protected when in a vulnerable situation. This can be done by running a fine wire to and fro across the ceiling then concealing and protecting it by papering over or skimming

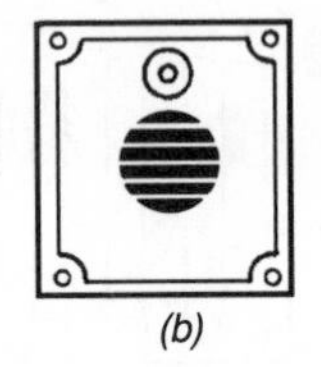

Figure 24 *(a) Small acoustic detector for fixing to glass. Responds only to sounds of breaking glass. (b) Non-contact acoustic detector detects sounds of breaking glass over 15 ft (4.5 m).*

with a ceiling finish. Even painting would be sufficient to retain the wire in position and protect it.

Special hard-drawn lacing wire is is made for this purpose or for similarly protecting door or partition panels. It is very brittle and snaps easily thus giving excellent protection, but it needs care in handling and must be protected by some form of covering.

As an alternative, fine copper wire, varnish-coated to prevent corrosion, can be used; the type used for transformer windings is particularly suitable. Although less brittle than the lacing wire, it breaks easily enough when subject to stress if of sufficiently fine gauge. As copper has a degree of elasticity, it should not break if subject to slight movements due to structural settling. Providing the connections are sound there is little possibility of false alarms.

With bungalows, the roof trapdoor should always be fitted with a sensor such as a magnetic switch, but the wiring must somehow be concealed from view both above and below. Alternatively, a pressure mat could be installed just below the trapdoor, or better still in addition to the trapdoor switch. Blocking the exit route, as discussed in an earlier chapter, is of particular importance here.

Acoustic detectors

We now come to a group of related detectors which are unlikely to find a use in the average home alarm system, being used principally for industrial installations. However, there could be some special situations for which they have an application.

The first of these is the acoustic detector. Also called *sonic detectors*, these are not to be confused with the ultrasonic devices described in the next chapter. The term embraces a range of sensors that operate mostly by the sound generated by breaking glass. Examples of the device are shown in Figure 24.

They consist of a microphone, an amplifier and an output relay. A special filter circuit is included which passes only those sound frequencies generated by breaking glass, and thus is immune from false alarms caused by other sounds. One type, also termed piezoelectric because the microphone is of the crystal contact variety, is mounted on the glass by means of an adhesive. The complete unit can be very small, little larger than 1 in (25 mm) in diameter. The filter eliminates low frequencies, passing those in the high 6000–8000 Hz range (piano top note is 4000 Hz).

Others are mounted on an adjacent wall or ceiling and have a normal air pressure microphone. Some of these have a more complex filter that responds first to certain low frequencies which are produced on impact, then to the high frequencies generated as the glass shatters. They will not trigger unless both the required frequencies occur in the correct sequence, thereby giving an even greater immunity against false alarms. This type should not be used for wired, laminated or toughened glass, however, as these glasses do not produce the correct frequency pattern.

The glass-mounted type must be positioned at least 2–3 in (50–75 mm) from any window frame because the intensity of vibrations in glass reduces from maximum at the centre of the pane to zero at the frame. Both types protect an approximately 10 ft (3 m) radius of glass. Some have an indicator light which latches on to show if it has been activated, or which one if there are several. If glass has been broken this may seem to be superfluous, but it does help to identify the source of a false alarm if one should occur.

They offer an advantage over window foil for large areas where a lot of foil may be considered to be visually detracting, and especially for multi-pane windows where there may be practical problems in foil laying. The disadvantage is that the visual deterrent effect of foil is lost. For multi-pane windows the space rather than the contact sensor should be used, as each pane would require its own contact detector.

A similar principle is used in wall sound detectors. With these, either structure or airborne sounds are picked up by the microphone and filtered so that only those produced by hammering, drilling and cutting activate the unit. Again it is mainly the high frequencies that are detected, so those produced by bumps, traffic noise and other normal happenings are ignored. A number of microphones can be connected to a single detector so that a large area or several separate areas can be monitored. The detector thus processes the output from all the microphones and open-circuits the loop if any are actuated.

As all acoustic detectors contain an amplifier, they require a power supply, and this is provided by the 12 volt auxiliary supply available in the control unit. Current consumption is 20–30 mA. An extra pair of wires is needed to convey this to the detector, so to include the anti-tamper loop and the detection loop, a six-wire cable is required.

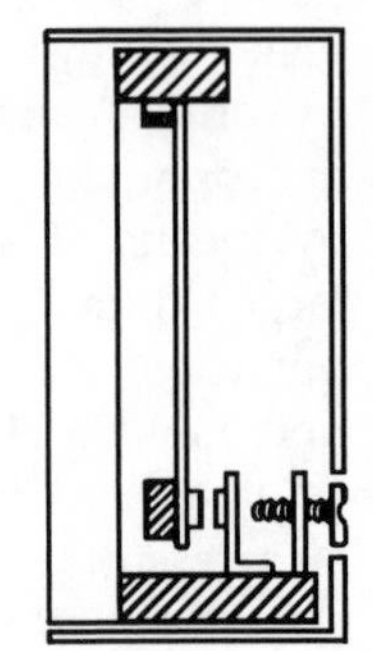

Figure 25 *Basic principle of vibration detector. Pendulum contact in close proximity to fixed contact. The latter is movable by screw to give adjustment of sensitivity.*

Vibration detectors

These are comparatively simple devices consisting of a leaf spring suspended at its top and having a weight fixed to its free end. Also at the free end is a contact that mates with another which is fixed to the case (Figure 25). The pressure between the contacts, which are normally closed, is adjusted by a set-screw which thereby sets the sensitivity of the device.

Any vibration or movement causes the contacts to part and so initiate an alarm. The detector can thus be used to protect windows as an alternative to the contact acoustic detector, or any wall or partition where forced intrusion is possible.

The main disadvantage is the difficulty of setting the sensitivity. If set too high, it may be affected by random vibrations. Passing heavy lorries in particular can make a large window vibrate strongly. There is no filter to eliminate inoffensive vibrations as there is with the acoustic detector. However, control boxes are designed to respond only to loop breaks longer than 0.2 seconds, so this helps to eliminate small vibrations which may just cause the contacts to part momentarily.

One advantage over the acoustic detector is that as plate glass is quite robust it may take several blows to break it. The acoustic detector is actuated only when the glass is broken, but the vibration sensor would almost certainly be triggered from the vibration of the first blow. With the alarm ringing after an unsuccessful first onslaught, the attacker would very likely be deterred from a further attempt, thus saving the loss of expensive glass.

Furthermore the vibration detector does not need a power supply, is cheaper, and requires less complex wiring. A large number could be used if required on walls, partitions or anywhere likely to be broken through. Its susceptibility to normal vibrations, though, would preclude its use in busy areas where there is much activity, and especially near main roads.

Impact detectors

Impact detectors respond to vibrations or impact as do the vibration detectors but they are more sophisticated. A piezoelectric (crystal) rod is supported at one end and any vibration causes the free end to oscillate and thus generate a voltage in the rod. A built-in circuit analyses the nature of the signal thus produced and actuates a relay if it exceeds a pre-set level and duration.

The device is thus less prone to false alarms than the simple vibration detector, and as it does not depend on gravity acting upon a weight, it can be mounted in any plane. It requires a 12 volt supply from the control box and is larger and more expensive than the vibration sensor. As each is self-contained with its own electronics, installing a number of them could be costly. It is therefore most suited for situations where only one or two are required. If more sensors are needed, it is more economical to use inertia detectors.

Inertia detectors

These sense disturbances and vibrations, but are particularly sensitive to low frequencies. Thus a gentle prising or levering of a structural member, or the motion generated by climbing a perimeter fence which may be ignored by a vibration or impact sensor, will trigger an inertia detector. As its name implies, the mobile component has a high mass to give it inertia, so that when the environment is disturbed, the housing moves but the component does not. Contacts are thereby broken, thus activating the alarm.

In one model, the component is a gold-plated ball seated on a pair of contacts, so forming a normally closed switch. Displacement of the ball from either contact breaks the circuit. A high-security version has in addition a closely spaced ring around the ball (Figure 26a). This serves as a normally open switch, so that any disturbance causes the ball to touch the ring as well as break contact with its supporting electrodes.

The maximum current that can be switched by this device is low because the area of the ball actually in contact with its supports is minute. A large current passed through such a small area could cause pitting and even welding to occur. The specified current is 0.2 mA, with an applied voltage not exceeding 2 volts. This is much lower than the 500 mA or so that the vibration sensor with its much larger contact area will permit. It is also lower than the current applied by many control units to the loop circuit. It thus needs its own control unit, which must not exceed the maximum current, and must process the result to distinguish between false alarms and intrusions.

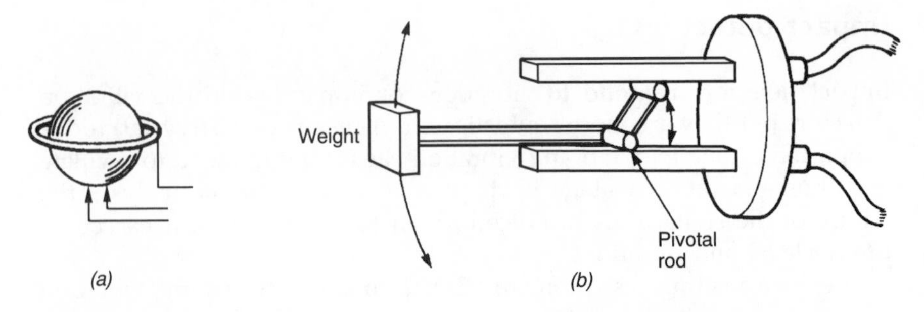

Figure 26 *(a) Inertia detector. A ball rests on two contacts forming a closed circuit. An extra ring contact surrounds the ball. It is used for detecting motion and vibration. (b) Inertia detector having a weight on the end of an arm which pivots on a lower contact when jolted, thereby breaking the contact against the upper contact.*

With another model, a lever having a weight at its free end is supported by a gold-plated rod which rests on a contact piece. An angled extension of the lever is terminated by another gold-plated rod that rests against the under-surface of a second contact piece above it (Figure 26b). The lever and rods are held against the terminal pieces by the downward pull of the weight, but any relative movement between weight and housing causes loss of contact.

This sensor too requires an analyser to control the current, set the sensitivity and sort out the true from the false alarms. It reacts if there is a series of small shocks, several medium shocks, or one large one.

With other analysers the response can be pre-set to a given number of impulses in a specified time. Below this, the alarm is not triggered, so random shocks against a perimeter fence by children playing, for example, have no effect. Some units have night/day changeover relays whereby the number of permitted shocks is reduced during the night when greater security is required and fewer random impulses are likely. Typical values are eight impulses in 30 seconds for daytime, and as few as four in 15 minutes at night. The values can be changed to give higher security (lower values) or greater immunity against false alarms (higher values) as circumstances dictate.

Inertia detectors are very sensitive but, in the event that a fence or structure is subject to continual or frequent small-scale vibrations such as from exposure to wind, magnetically damped sensors can be used. A similar effect may be obtained by reducing the sensitivity control. The frequency response is generally from 10 to 1500 Hz and so covers the region generated by climbing or attack.

Inertia sensors are the best method of protecting fences, but they need to be placed at about 10 ft (3 m) intervals. This could mean quite a number if a large enclosed area has to be protected. However, a dozen

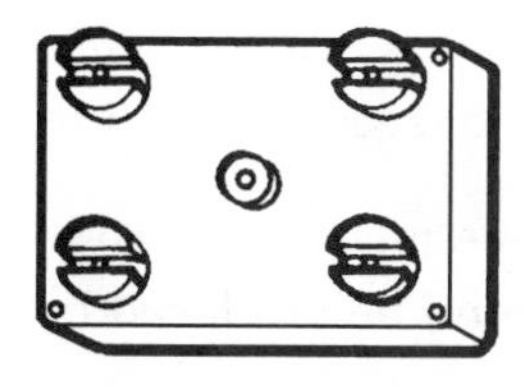

Figure 27 *Safe limpet containing vibration and magnetic switch.*

or more can be connected to a single analyser and it is the analyser that is the most expensive item. Some analysers have separate zones which enable different sensitivities to be set for the respective groups of sensors. This is useful where sensors are used on different surfaces or positions. Also, indicators that latch on show which region produced the alarm.

Like the vibration detector they can also be used to protect large areas of glass, but their rejection of inoffensive vibrations makes them more suitable for this task in active environments such as by main roads. However, even with these, the sensitivity setting is very much a case of trial and error. The best method is to set the detector for high sensitivity during the daytime, but with the analyser disconnected from the main alarm system. The latching indicator will probably soon show an alarm condition. Slightly reduce the sensitivity and leave for a further period, continuing the process until no alarm is shown at any time. Leave it at this setting for a few days to make sure a casual vibration or shock will not trigger it before connecting to the main system.

Inertia detectors can be used to protect walls and partitions and the spacing should be 12–16 ft (4–5 m). Here they should be mounted at least 3 ft (1 m) away from floors, ceilings or abutting walls. For ceilings and roofs the spacing can be 20 ft (6 m) and the sensors should be mounted on beams and rafters. Some units are made for flush mounting into structural members while others are intended for surface mounting. In every case they must be positioned vertically and there is usually some mark on the casing to show the correct orientation.

Safe limpets

Not every home sports a safe, but for those that do there are sensors that will warn of any attempted interference with it. These could be used on filing cabinets or computer equipment consoles. Several types are obtainable, but most are fitted magnetically to the unit being protected (Figure 27).

The simplest type contains a vibration detector which operates if any attempt is made to force the safe open. It also has a contact which is held in place magnetically for as long as the unit is fixed to the safe. If it is removed, the contact opens and the alarm is triggered. Another type utilizes an inertia sensor which responds to low frequencies as well as the higher ones.

A third type includes a thermal switch as well as vibration contacts. This detects an abnormal rise in temperature as would be produced by a thermal lance or other metal cutting equipment.

Summary

It can be seen from the foregoing that there is a considerable armoury available in the form of various types of detectors. Several different ones, as we have seen, can be used for one specific application, while some are suitable for many. Each has particular features that make it the best choice in a given situation.

All except the pressure mat have one thing in common: they are perimeter detectors, that is they protect the outside boundaries of the premises and prevent intruders from getting inside without sounding the alarm. They thus complement the physical security which should be the first line of defence, as we saw in Chapter 1.

It is always wise though to have a second line of defence just in case the first one is somehow penetrated. Also it is sometimes impossible to effectively protect the whole perimeter. The next chapter deals with further devices which are generally known as space protectors.

6 Space protectors

The detection devices described in the previous chapter were all (with the exception of pressure mats) operated by attempted penetration and movement of some part of the perimeter of the premises. For this reason they are often described as perimeter defences. It is essential wherever possible that the perimeter be fully protected, as it is far better to keep intruders out than to detect them once they are inside.

In some cases full protection of all parts of the perimeter is impractical. Even where it is provided for high security, there is always the possibility that it could be breached by a determined and knowledgeable intruder. As a back-up, space protection detectors, often called *volumetric sensors*, can be added to the system to cover this need. They should not, though, be relied on as the sole means of defence. One reason is the fairly high current of over 25 mA taken which could not be sustained for long by ordinary standby batteries if the mains should fail. Other reasons we shall see.

The main types of space protector are ultrasonic, microwave, active infrared and passive infrared. There are also some lesser known ones that have special applications. The ultrasonic and microwave detectors rely on the Doppler effect, so we will firstly describe exactly what this is.

Doppler effect

The effect itself is quite familiar: the siren of an approaching ambulance or police car sounds higher in pitch than it actually is, yet when the vehicle passes and moves away the pitch sounds lower. This is due to the Doppler effect.

The pitch of a tone is governed by the number of sound pressure waves that reach the listener per second. This rate is termed the *frequency*, the unit of which is the hertz, abbreviation Hz (a common multiple is the kilohertz, abbreviation kHz, which is equal to 1000 Hz). When either the source or the listener approaches the other, the waves come at a faster rate, so the frequency is higher. It is rather like swimming out to sea: you encounter more waves per minute than you would standing in the water waiting for them to come to you.

If the source and the listener are separating, then we have an opposite effect. The rate is decreased, just as it would if you swim or surf back to the shore; you may even encounter only one wave that carries you back.

Pitch thus depends on relative movement between source and receiver. If an object reflects waves to a receiver, it in effect becomes a source, and if it moves towards or away from the receiver, a frequency difference will be perceived. This is the principle used to detect movement of an object in the protected space; waves from a source are reflected off any object that comes within range, and are picked up by the receiver.

Ultrasonic detectors

The term *ultrasonic* denotes sound frequencies that are above the range of human hearing, the upper limit of which is 16 kHz in healthy people in their twenties. The frequencies used in these detectors vary between different models, ranging from 23 to 40 kHz. The frequency is generated by an electronic oscillator and is fed to one or more loudspeakers. These need to be very small, smaller in fact than the tweeter in a hi-fi loudspeaker, because the moving parts must move very rapidly and so their mass must be kept to a minimum. This is an advantage for an alarm system as the units can then be made small and unobtrusive. The same unit contains a microphone with amplifying and processing circuits.

Ultra-high-frequency sound is thus projected into the protected area. Some of it is received directly by the microphone from the loudspeaker, while some is picked up after being reflected from walls and objects in the room. When there is no movement, both direct and reflected sounds are of the same frequency. Should movement of an object occur, the sound reflected from it will undergo a change of frequency owing to the Doppler effect. Thus the microphone now picks up two frequencies: the original received directly from the loudspeaker, and the shifted frequency reflected from the moving object.

If two different frequencies are mixed, the result is that a third one appears that is the difference between the two. When the two frequencies are close, the third is often termed a *beat note*. The effect can sometimes be heard when two internal combustion engines are idling at slightly different speeds: the beat note is heard as a throbbing sound that varies in frequency as the speed of one or other of the engines changes. (Older readers may remember the throb of twin-engined German aircraft during the war.)

The processing circuits detect the presence of this beat note and actuate the output relay which is of the normally closed type. The device can thus be connected into a normal loop, but it is preferable for space detectors to be connected to a separate zone from the perimeter sensors so that rapid identification of the alarm source can be made. A power supply of 25–50 mA is required from the control box.

Ultrasonic detectors usually have a range up to 30 ft (9 m), but they

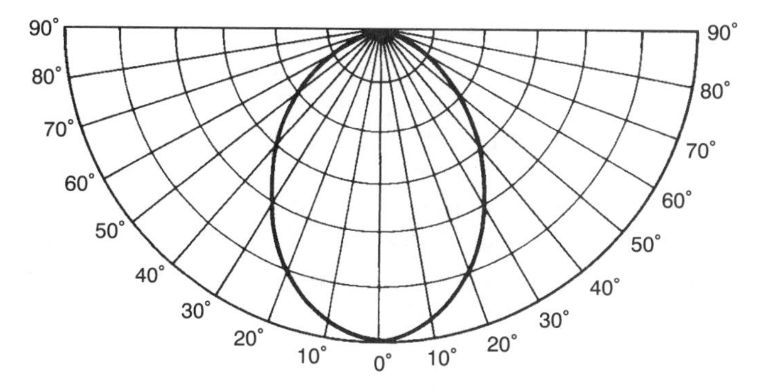

Figure 28 *Polar diagram of the response of an ultrasonic detector.*

have a sensitivity control which can reduce the range down to about 10 ft (3 m). The polar pattern is in the form of a narrow lobe (Figure 28) and so the device affords little protection at the sides. It should therefore be pointed towards the expected point of entry. As the Doppler effect is greatest for forward or backward motion, the sensor is more sensitive to these than to side-to-side movements. This is another reason to mount it facing the possible entry point.

Mounted in this position, the device is virtually impossible to defeat as any movement towards it, however slow and deliberate, will set it off. Anti-tamper connections are usual as is also a microswitch which is released if the casing is removed, but these are not usually used in domestic situations. Units are often disguised as hi-fi speakers, office intercoms, and even books.

The main disadvantage is susceptibility to false alarms. These can originate in several ways. Ultrasonic sounds can be produced by vehicle brakes, gas or water jets, television sets and possibly computer VDUs and printers, among others. In industrial environments it is possible for leaking compressed air lines and almost any machinery to generate these frequencies as harmonics of their normal noise output. Such sounds will be picked up along with the sensor frequency and so produce a beat note which it will interpret as a Doppler product.

Another source of false alarms is air turbulence which can distort the sound propagation and produce Doppler shifts. This will result from forced air central heating systems, but can also be caused by other heating installations. Draughts, fans, moving curtains and the like are other possible causes.

The sensitivity control helps as this can be set to give only the necessary range. Reducing it to the minimum required reduces susceptibility to false alarms.

Modern ultrasonic detectors are somewhat less prone to these ills than the early models, as they have built-in circuitry for minimizing them. One type uses an averaging circuit which only responds to a net change in target distance. Vibrating objects, swaying curtains and some air turbulence average out to a zero position change, and so do not trigger the alarm. However, even with these, care should be taken in their use. It would be unwise to use them if any of the above-mentioned factors was present. Because of these problems they are less used now than they once were, especially as better space protectors are now available.

Microwave detectors

These use the Doppler effect and work in a similar way to the ultrasonic detector. The difference is that they employ radio waves that oscillate at extremely high frequencies. The standard is 10.7 gigahertz (GHz: 1 GHz = 1,000,000 kHz), but 1.48 GHz is also employed.

Receiver and transmitter are in the same unit, and the receiver receives some direct radiation from the transmitter along with that reflected from objects within the protected area. If the reflected frequency is shifted by the Doppler effect, a beat frequency is produced and the result actuates the alarm circuits.

A feature of microwaves is that they will penetrate wood, glass, plaster and, to a limited extent, brick. Microwave detectors also have a much greater range than ultrasonic sensors, and will cover up to 150 ft (45 m). They are thus well suited for the protection of large areas, which is why they are often used in warehouses, not only because of the size, but also because large stacks of cases and goods could afford shelter to an intruder from other types of volumetric detection. As microwaves pass through these, there is no shelter from them. Propagation patterns are usually tear-shaped in the horizontal field, with a narrow lobe in the vertical plane. Some units are designed with long narrow horizontal lobes to cover similar-shaped areas, or have split beams to give V-shaped coverage. Deflector plates can modify the pattern.

Steel cabinets, shelving or machinery could provide cover, however, because metals reflect microwaves. A solution to this problem is to mount the unit in the centre of the ceiling as shown in Figure 29. Some models are designed for this purpose and have a hemispherical distribution pattern. The feature of these is that they can see over metal obstructions, so there is little cover for any intruder.

Coming from this angle the microwaves do not need to penetrate 'soft' materials such as wood or plastic and so can be of a lower frequency such as 1.48 GHz. The range of a ceiling-mounted unit at a height of 10 ft (3 m) is a circle at floor level of some 60 ft (18 m) diameter.

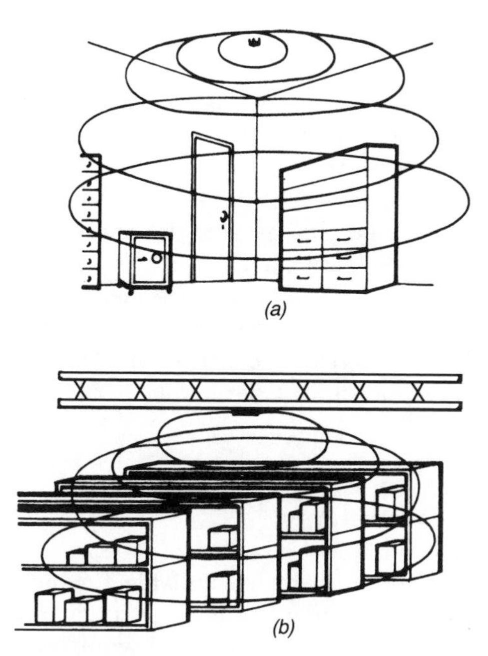

Figure 29 *Ceiling-mounted microwave unit gives good coverage over limited floor area. There is virtually no penetration outside the area except through the floor. Ideal for ground floor areas. (a) Typical small office area protection. (b) Excellent coverage in factory storage area. Steel racking gives intruder shelter from horizontal beams, but none from overhead radiation.*

Microwave sensors are not as subject to false alarms as ultrasonic detectors. Draughts, air disturbances or turbulence have no effect, nor does any type of sound wave. Fluorescent and neon lights can generate radio signals that could be accepted by the receiver, but there are usually filters that remove these electronically.

The main possibility of false alarms comes from the penetration of microwaves beyond the perimeter walls or, in the case of the ceiling-mounted units, penetration of the floor to the room beneath. To reduce this possibility, the sensitivity must be adjusted so that the range extends only to the perimeter boundary. Even then it is possible for some overspill, especially through windows or doors. The effect can be avoided by curtailing the sensitivity to range somewhere just short of the boundary, although this would reduce the level of security.

This really is the problem with most volumetric systems, and is the reason why they should be used as a back-up rather than the sole means of detection. If the perimeter is well protected, there need be no qualms about restricting the range to prevent false alarms.

As with the ultrasonic sensor, there is no way the microwave detector can be defeated when it is switched on as any approach immediately triggers it.

The units are usually totally enclosed in a plastic box without any apertures, as the microwaves travel through the plastic. So it is not easy to tell just what the box contains; it could be merely a junction box. This anonymity is obviously a feature in its defence. Some ceiling models are flush fitting and so give no clue as to their true nature.

For applications where very high security is required, a microwave monitor can be installed. This can be positioned remote from the unit, but within its range. It detects the presence of microwaves, so if they are absent for any reason, it gives an alarm signal. This would also protect against inadvertent shielding by a large metal object left in the path of the microwaves.

Microwave beam-breaking

Microwave Doppler systems are not suitable for outdoor sites as the perimeter is often a wire or wooden fence which is quite transparent to microwaves: any moving object just beyond would activate the system. Furthermore birds, cats, foxes or other animals encroaching on the protected area would trigger an alarm. Reducing sensitivity to fall short of the perimeter would remove protection just where it was needed.

Another type of microwave system does not use the Doppler effect. Instead it transmits a narrow beam to a receiver situated at a remote point and thereby protects the space between. The alarm is triggered if anything breaks the beam. The range can be considerable, up to 500 ft (150 m), and the device is very suitable for guarding outdoor sites such as large gardens, grounds, etc.

Although the detector is a space protector, the space protected is a straight line and so the system is not volumetric. It is necessary for each boundary to have its own transmitter and receiver. This arrangement could guard up to 2000 ft (600 m) of boundary fencing, but four sets are required for protection on all sides. Even so, this is far more economical than using inertia sensors every 10 ft (3 m). The device has an inherent anti-tamper property in as much as any interference with the transmitter prevents the beam from reaching the receiver.

The beam is not clearly and sharply defined like a focused light beam, but is rather diffuse. Its divergence can produce a beam width of up to 20 ft (6 m) at full range. Two detectors could thus be used spaced horizontally looking at the same transmitter, to give a double line of protection.

The current required is rather high, 150–250 mA for each detector, so to protect a site with four units takes about 1 A. This is more than

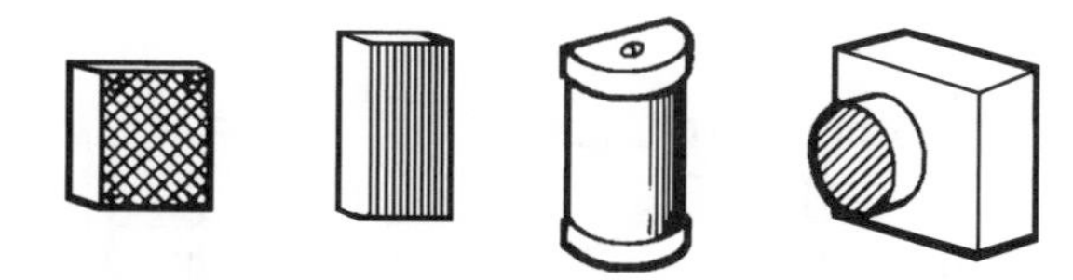

Figure 30 *Selection of infra-red units. Projectors, receivers and mirrors are housed in identical cases. Dummy cases are also available to further confuse intruders.*

usually delivered by control boxes and so a separate power unit is required. A standby battery would need to have at least the capacity of a car battery, which would last about 80 hours before recharging.

Active infra-red detectors

These operate in a similar way to the microwave devices in that a generator radiates a beam which is picked up by the receiver. If the beam is broken by an intruder the alarm is triggered. The infra-red rays are usually produced by a gallium arsenide crystal and hence they are sometimes called gallium arsenide rays, but they can be generated by a wide variety of means.

Any source of heat radiation will usually produce infra-red radiation because both lie just below visible light, in the same part of the electromagnetic spectrum. Thus lamp bulbs, heaters and even a pocket torch generate both to a greater or lesser extent. This opens the possibility of an intruder defeating an infra-red system by directing a portable infra-red source at the receiver as he moves across and breaks the original beam.

This type of interference is prevented by pulsing the beam. The frequency of the pulses varies, but 200 pulses a second is common. The receiver only responds to a pulsed beam of the same frequency; if the pulses are replaced by a steady beam, it signals an alarm.

Projector and receiver are housed in identical cases (Figure 30); hence it is not possible to identify which is which. This adds difficulty to anyone trying to defeat the system, and the problem is compounded by the use of additional dummy housings which are available for the purpose.

The only way to evade the ray is by avoiding it, such as by crawling underneath it. To do this the intruder must be aware of its presence and its path, but even so evasion becomes impossible if the beam is laced across a vulnerable point in zigzag fashion. This can be done by means of mirrors; unlike the microwave beam which is too diffuse, the infra-red ray can be reflected at several points by small reflectors just like rays of visible light (Figure 31).

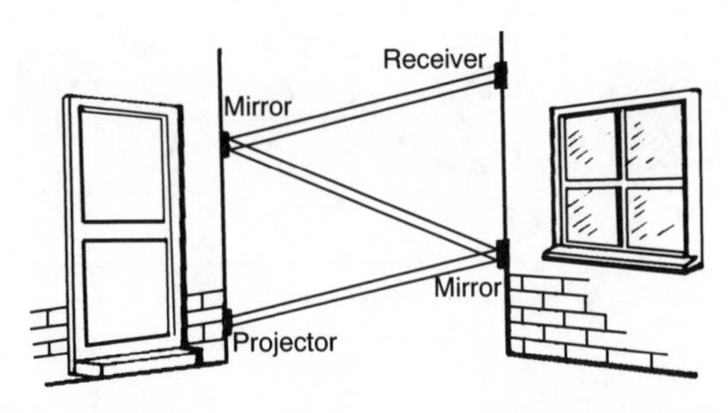

Figure 31 *Lacing an infra-red beam across an area to make avoidance difficult. Lower units should be closely spaced to prevent intruders stepping through or crawling under the beams. False alarms are possible from wildlife etc.; to avoid this, use two parallel beams.*

The range varies from 65 ft (20 m) for the indoor type up to 1000 ft (300 m) for the most powerful outdoor type. When reflected from a mirror the ray is attenuated and so the range is shortened. For a single reflection the range is reduced to 75 per cent; a second reflection reduces the range to 56 per cent, and a third to 42 per cent. It is not recommended that more than three mirrors be used.

The rays will also penetrate clear glass just like light and so protection can be extended through partitions and windows. Here again there is a reduction in range for each glass pane although the reduction is less than for mirror reflections. For 2 mm window glass the range is reduced to 84 per cent for a single sheet; for a second sheet the range becomes 70 per cent, a third reduces it to 60 per cent, while a fourth drops it to 50 per cent.

An infra-red beam is thus an alternative to the microwave for four-sided perimeter external protection, and has the advantage that only a single projector and receiver need be used with mirrors to divert the beam around the area instead of four microwave sets. Furthermore, the projector and receiver are in the same corner, thereby saving wiring.

There is a possibility of false alarms due to birds or windborne objects interrupting the beam. This can be avoided by running two parallel beams one above the other and connecting the receiver output contacts in parallel so that both beams must be broken simultaneously to produce an alarm. This requires two sets, but there is a dual-beam model that serves this function with a single unit.

Another solution is the *double-knock analyser*. This is a circuit that can be wired to almost any sensor that could be subject to false alarms. It triggers an alarm only after two alarm conditions have been received within a set time that can be adjusted from 10 to 80 seconds. Security is slightly lessened by this, but it greatly reduces the chance of false alarms.

A disadvantage with external infra-red beams is that they are attenuated

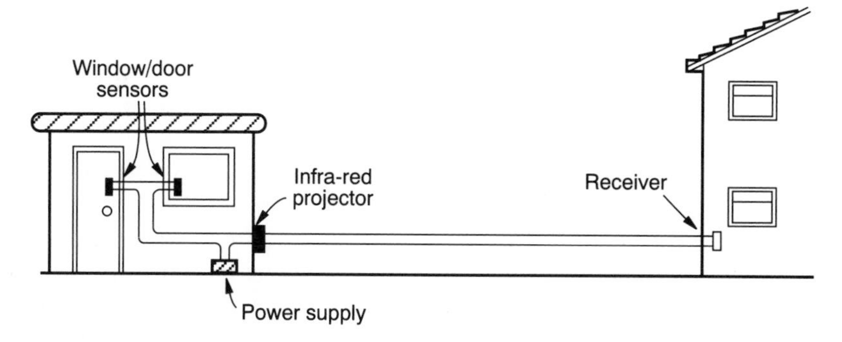

Figure 32 *Separate alarm system in outbuilding can be tied into the main alarm by means of infra-red beam. Sensor loop is in series with power supply to projector, so if sensors are actuated the projector stops. Receiver on main building then sounds the alarm. Latching circuit is not necessary as this is included in main alarm, so alarm continues even if beam cessation is only momentary. Protection over intervening area is also achieved.*

by fog and rain. Also condensation on the lenses of projector or receiver reduces the transmission through them. It is customary to include a heater in both units to keep them free from condensation, and this takes extra current from the supply. Heaters are not built in to units intended for internal use and this is their main difference.

The mirrors used to deflect the beam are passive and have no power supplies. Thus they have no heaters and so can be affected by condensation or raindrops and ice and snow collecting on the surface. This is a serious drawback which could result in false alarms. The use of mirrors outdoors to achieve perimeter protection should be confined to locations where fog, ice and snow are rare, and even then the total distance should be well within the range of the unit to allow a generous safety margin.

An infra-red beam can be used to link an auxiliary alarm system in an outbuilding to the main system in the house, as well as to protect the intervening space. Figure 32 shows how this can be done. The sensors in the outbuilding are in series with a loop that carries the power supply for the infra-red projector. If they are actuated, the projector is switched off and the receiver on the main building signals an alarm to the main system. If the beam is broken, this also triggers the alarm. It should be noted that microswitches rather than magnetic switches should be used for sensors, as the latter would not carry the current required to supply the infra-red projector.

Where large areas are to be protected, the active infra-red system is a useful option. For smaller areas and especially indoors, a better alternative is the passive infra-red detector.

Passive infra-red detectors

The passive infra-red detector (PIR) is a newcomer to the security scene compared with the types we have so far discussed. It has been made possible by the development of highly sensitive ceramic infra-red detectors. The device does not have a projector, but operates by detecting the infra-red radiation from a human body.

The received radiation is focused on to the infra-red sensor by either a curved mirror behind it or a curved plastic lens placed in front. The curve is not continuous but is broken up into a series of vertical facets which have a particular function. The detector has two sensitive areas, and some facets focus on to one while others focus on to the other. When a heat-emitting object moves across the field of view, the image thus appears alternately on these two areas. An electrical output is generated only when there is a varying difference between them; nothing happens when either, neither, or both receive steady radiation. Thus only a moving object produces an output; a stationary source of heat has no effect. This serves to eliminate many possible false alarms.

The facets produce detection zones like spreading fingers, and it is the crossing of these zones that produces the required fluctuations at the detector. There are usually two or sometimes three sets of horizontal divisions as well as vertical ones in the mirror or lens. These produce a second or third array of detection zones, one above the other and at different planes and angles. These lower planes give protection to the space underneath the main zones and closer to the sensor. The effect is shown in Figure 33, which illustrates eight different PIR detection characteristics, with ranges, plan (overhead) views and side views.

Detection is almost instantaneous and the range is surprisingly long considering the device relies on body heat. The usual range is 40 ft (12 m), but many have a range of 50 ft (15 m). Some special ones with a narrow field designed for corridors have a range of up to 130 ft (40 m). The normal field is fan-shaped with a 90° spread, which is ideal for corner mounting.

When the unit is mounted vertically on a wall at a height of about 6–8 ft (2–2.5 m), the detection zones are angled downward so that the highest ones meet the ground at the rated range. A shorter range may be required such as for covering a yard bordered by railings. With small areas bounded by an open perimeter such as railings, the range could be too long, and false alarms could be generated by movement on the other side. A sensitivity control is usually included for range shortening; more reliably, the device could be tilted downward so that the detection zones are shortened.

It may be required to give protection closer to the unit in preference to at a distance, and here again downward angling will achieve it by

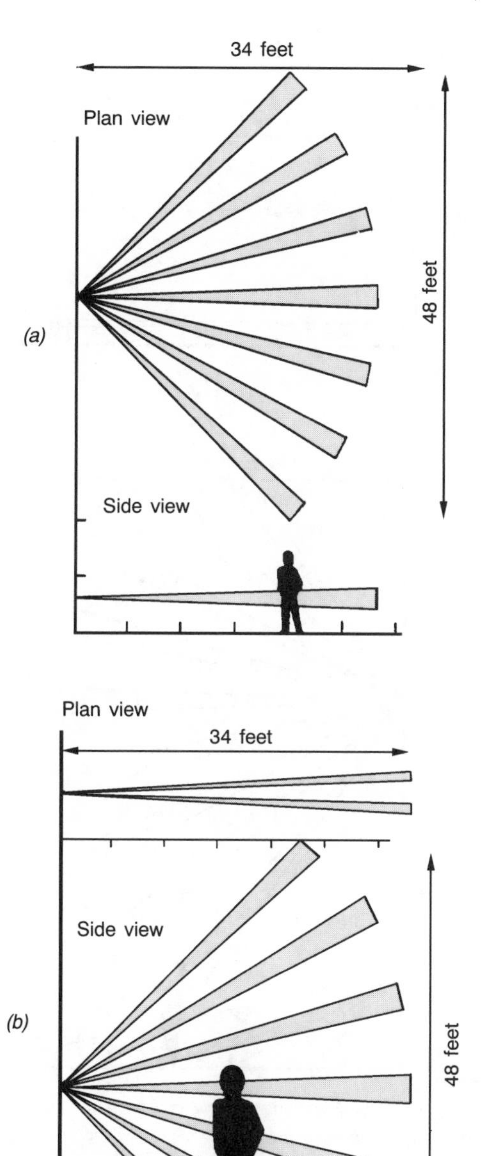

Figure 33 *PIR sensor patterns and ranges. (a) PIR designed to avoid triggering by small animals. (b) PIR with curtain effect to give large vertical cover.*

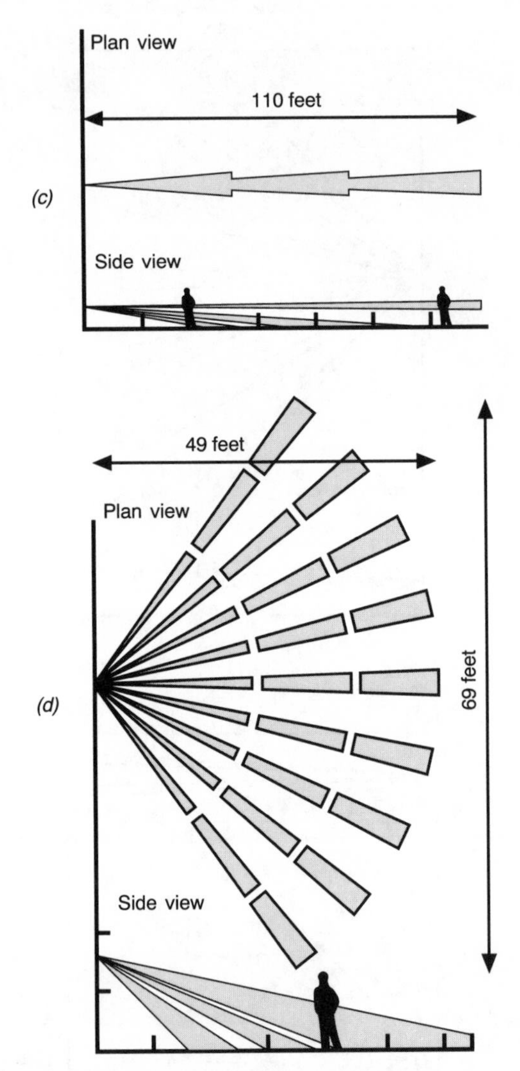

Figure 33 *(continued)* *(c) PIR with long narrow coverage for corridors and stairways. (d) Large-area PIR with three vertical zones.*

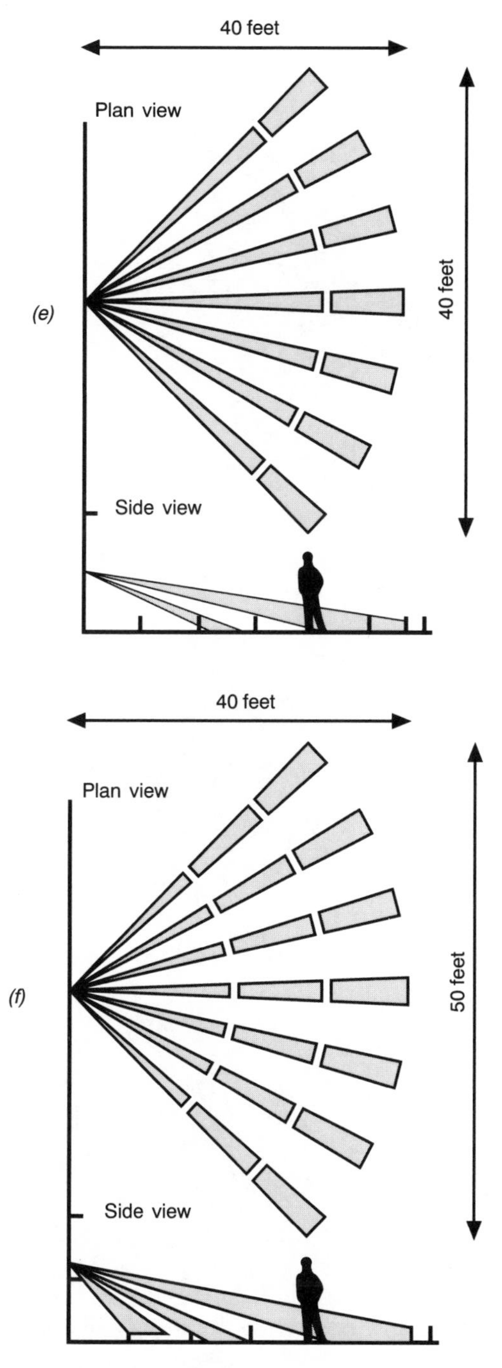

Figure 33 *(continued) (e) Medium-range general purpose PIR. (f) Medium-range PIR with wide coverage area at a maximum range and a close zone for nearby detection.*

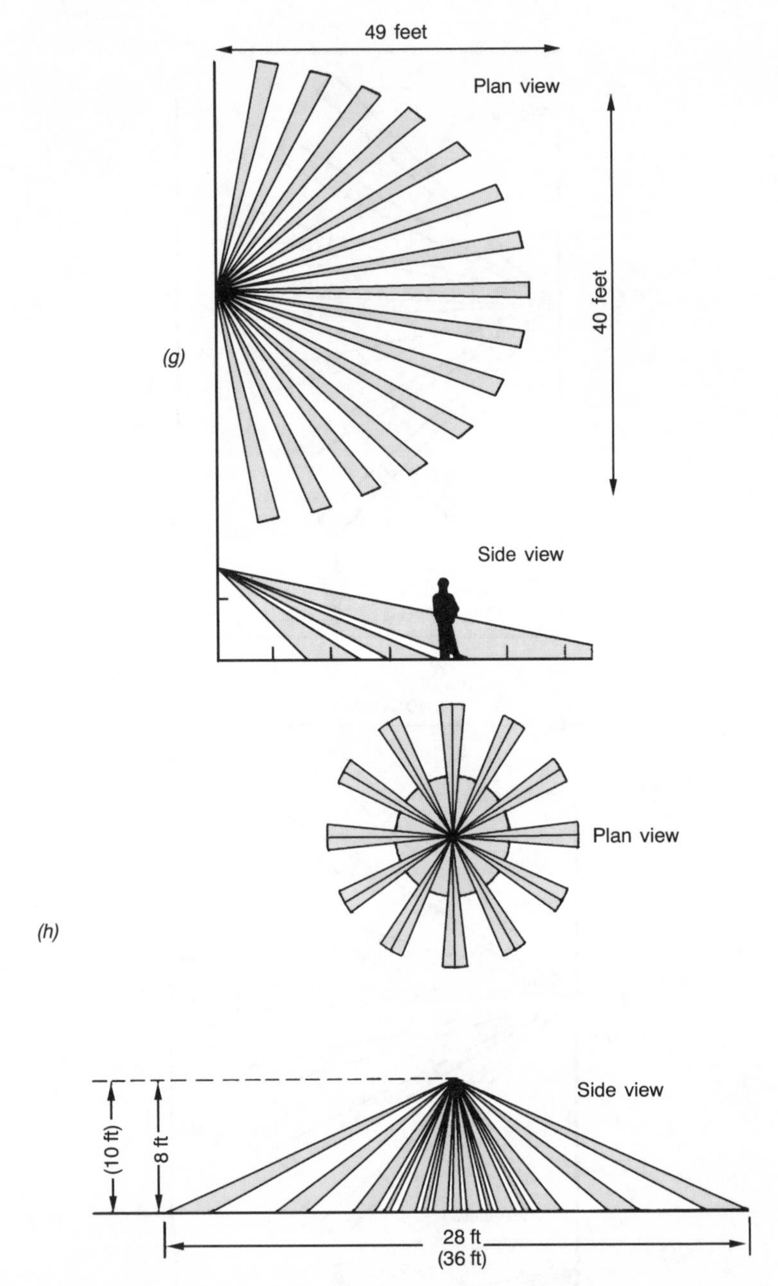

Figure 33 *(continued)* *(g) Wide-angle PIR for mounting on flat wall instead of usual corner position. (h) Overhead PIR gives wide circular coverage dependent on height.*

bringing the lower zones closer. Conversely, a slightly longer range than that specified may be obtained by mounting the unit higher, so that the upper zones meet the ground at a greater distance. In this case the sensitivity control would have to be set to maximum.

False alarms due to inanimate moving objects are not possible unless the object is warm because the device is actuated only by objects that are both warm and moving. The detector is not affected by heating systems, with the possible exception of forced air heating under certain conditions.

While a strong infra-red ray from a projector will penetrate glass, as we have already seen, infra-red radiation from a human body is insufficient to activate a detector after passing through glass. So there is no danger of false alarms due to someone passing the other side of a window in a protected area.

Where random momentary triggering may be possible, models are available that have double-knock capabilities. They ignore a set number of disturbances, which can be two to four, and trigger on the next. An intruder would certainly cross more than two zone boundaries, so there is little loss of security.

The most common causes of false alarms in outdoor applications are animals. Cats are no respecters of property boundaries, and will, if entering the range of a PIR, set it off. At least one model has only one plane of detection zones, and that is aimed horizontally with no downward tilt. It could be described as a moggy-proof model, but needs to be mounted at no more than waist height to ensure it does not miss human intruders.

The response of a PIR to animals suggests a possible use for poultry keepers in detecting foxes. For such an application the device should be mounted low but tilted back to achieve the range. Alternatively the moggy-proof unit could be mounted just above ground level. Human marauders would also be caught by it.

The PIR unit output is a relay with normally-closed contacts – that is, they are closed when on guard but open when triggered or when the power supply is removed. Connection is thus made to a normal loop, and a power supply is required of 12–25 mA. Anti-tamper connections are also provided which can be ignored for domestic installations. A four-wire cable is thus required for loop and power supply.

Resetting occurs as soon as the cause of the alarm has been removed, although of course the control box will latch on in the alarm condition. Some models have a latching facility. With these an extra wire back to the control panel is needed, requiring five-core cable. An indicator light on the unit remains on and the relay contacts stay open after the cause of the alarm has been removed. By this means the unit initiating the alarm can be identified when two or more PIRs are used on the same control-box zone.

PIR detectors are available not only as sensors linked to the main security system but as auxiliary detectors, and most outdoor sensors are in this form. These are combined with floodlights and an automatic daylight sensor switch. The floodlights come on when a person or a vehicle comes into range, and the daylight switch ensures that the device is not activated in daylight.

With these, the device latches on and the lights remain lit for a period after the activating cause has ceased. The period can be pre-set from several seconds to many minutes. After that the unit resets and is ready for further triggering. If a body is still present and moving in the detection area, the light remains on.

While floodlighting is a strong deterrent to intruders, it can be expensive to run all through the hours of darkness at a sufficient level of illumination. Lighting that comes on only when someone approaches is economical and also has an unnerving effect on an intending intruder. For legitimate callers it is very convenient to have the outside of the premises illuminated on approach without the need for manual switching, and this also serves as a deterrent against personal attack. Different parts of the outside area can be illuminated by separate units as they switch on and off independently.

It can be seen from this that the passive infrared detector has many advantages over most other types of space protection. It has a high immunity from false alarms; it has a wider spread than ultrasonic or microwave detectors; it has a longer range than ultrasonic devices; it can be used outdoors; and being based on optical principles, it can be angled to give exactly the coverage required, and accurately avoid adjacent unprotected areas. However, these detectors are not totally defeat-proof. It is possible to evade them with a heat-shield made of aluminium foil or a photographer's silver parasol, but these are very unlikely accessories for the average burglar!

One characteristic of all PIR detectors is that they take a few minutes to stabilize after the power is switched on and so are not immediately on guard. Once they have stabilized, the response to intrusion is instantaneous unless a double-knock mode is operating.

Electrostatic devices

Unusual and rarely encountered, the electrostatic sensor uses the same principle as that occasionally seen in shop window advertising displays. In these a mobile display is connected to an electrode which is stuck on to the inside of the glass. Placing a finger over the electrode on the outside introduces a capacitance to earth because the human body is virtually an earthed mass. This capacitance affects a delicately balanced electronic circuit which trips a relay and switches on the model.

Table 1 *Motion detector sensor characteristics*

Condition	*Ultrasonic*	*Microwave*	*Passive infra-red*	*Active infra-red*
Draughts, turbulence	×	✓	+	✓
High-pitched sounds	×	✓	✓	✓
Heaters	×	✓	+	✓
Moving curtains etc.	*	+	✓	✓
High humidity	+	✓	+	+
High temperature	+	✓	*	✓
Static reflections	✓	*	✓	✓
Sunlight	✓	✓	+	+
Movements beyond perimeter	✓	*	✓	✓
Vibration	*	*	✓	✓
Water in plastic pipes	✓	*	✓	✓
Small animals	*	*	×	+
Mutual interference	+	+	✓	✓
Long range	–	✓	–	✓
Wide coverage area	–	✓	✓	–

✓ No problems.
+ Slight problems under certain circumstances.
* More serious problems; needs careful setting up and siting.
× Major problems; not recommended with this condition.

In the case of the alarm, an electrostatic field is generated around the sensor. Any mass introduced within the field upsets the balance and the associated electronic circuit triggers the alarm. The range of the device is limited to a few feet which, when compared with other systems, is a disadvantage. However, it can only be actuated by a mass of predetermined size, and so cannot be triggered by any of the usual causes of false alarms. It could be used as a back-up to protect limited areas such as in front of safes, valuable paintings or trophy cases.

Summary

For indoor space protection the choice lies between ultrasonic, microwave, active infrared and passive infrared sensors: Table 1 summarizes their characteristics. The electrostatic system is a rarity and can be dismissed other than for a third line of defence behind one of the others where very high security is necessary.

Ultrasonic detectors were once the main choice for space protection but have fallen from favour since the appearance of other types, especially the passive infrared. Their main disadvantage is their susceptibility to sources of false alarms. They can be triggered by noises having harmonics in the ultrasonic range. Also they can be activated by draughts, by innocuous moving objects such as curtains, and by air movement generated by heating systems. While some units have a degree of immunity to these by only responding to net movement, there is still a higher risk of false alarms. Another drawback is their range, which at around 30 ft (9 m) is less than microwave or passive infra-red detectors. Furthermore, the sideways coverage is restricted to a narrow lobe.

As with the microwave detector, the use of the Doppler principle gives the device a high sensitivity to movement towards or away from the detector. It is less sensitive to sideways movement. The passive infra-red sensor has the opposite characteristic, being more sensitive to sideways movement. Thus an ultrasonic detector gives good protection in corridors and other long narrow areas where the intruder would be approaching the unit.

As it responds to movement by any object, the ultrasonic detector is virtually impossible to defeat. In theory a passive infrared detector could be evaded by an intruder carrying a large heat-shield, although this is very unlikely. So the ultrasonic detector can give a theoretically higher degree of security than the PIR, but the liability to false alarms is a big disadvantage.

Microwave systems are not susceptible to the interference sources that afflict ultrasonic units, but they have other disadvantages. The main one is penetration of boundaries, especially through doors or windows. They can thus be triggered by movement outside the perimeter unless the sensitivity is carefully adjusted.

Solid-object penetration can be an advantage in places such as storage buildings where there could be many large objects to provide cover for an intruder from other types of detector. In this situation, ceiling mounting is the best, and this also eliminates boundary penetration other than through the floor. The range obtainable from microwave detectors is greater than that offered by ultrasonic devices.

Active infra-red systems, although included in this section, are not volumetric protection devices as they guard only a narrow area through which the projected beam travels. Most of the possible applications are better served by passive infra-red detectors, but there are some for which the beam system could be preferred.

Unlike the passive infra-red sensor, the active device is effective through several layers of glass, and so could be used to protect through glass. Furthermore it can be used to protect items at private art displays and exhibitions, where there are boundaries which the public can approach to view but cannot exceed; or anywhere that invisible boundary protection is required.

The passive infra-red detector has taken the place of most of the other types for the majority of applications, although there are a few, such as those mentioned above, for which others are better. Responding only to objects that both move and generate infra-red radiation eliminates many causes of the false alarms in other sensors. A wide angle, normally 90°, and a range of up to 50 ft (15 m), give protection over a much larger area than most. It can be precisely aimed to give just the coverage required and so not encroach on outside areas.

Although most sensitive to sideways motion, it is by no means insensitive to forward and backward movement. As the detection zones spread out fanwise from the unit, and most detectors have two or three vertical planes of detection, it is virtually impossible to move more than a few inches within range without traversing across two or more zones.

While the possibility of avoiding detection by using a heat-shield has been suggested, it is a remote one. It would mean that the intruder would have to know that PIR detectors were installed and their position, would need knowledge of how they work, and would have to come equipped with a shield that was large enough, light and manoeuvrable. While this is quite possible for the professional thief after a high prize, it is very unlikely for the casual opportunist burglar. Even so, where very high security is required, another system such as microwave or ultrasonic should be used as well as PIRs.

There are some dual-technology units that combine a PIR detector with either a pair of microwave detectors or an ultrasonic detector. Some of these can be set either for *both* sensors to be actuated in order to trigger an alarm, or for *either* of them. In the first mode any condition which may produce a false alarm in the one may not affect the other, so conferring a high immunity against false alarms, which is useful in difficult environments. In the second mode, an intrusion missed by one could be detected by the other, thereby increasing the security. The choice can thus be made according to circumstances. One unit combines a PIR with an acoustic breaking-glass detector, but in this case they act independently.

For outdoor applications the choice is narrowed to microwave, active infrared and passive infra-red. To protect a large perimeter fence, four microwave beam detectors are required. A single infra-red beam could do the same job using deflection mirrors, but there are snags. Firstly, although the infra-red detector has a range of up to 1000 ft (300 m), this drops considerably if the beam is reflected three times. The range is actually reduced to 400 ft (120 m). So, this gives a rule of thumb that to enclose four sides of a perimeter with a single beam reduces the range to 40 per cent of the original.

However, to avoid false alarms due to small objects breaking the beam, either a twin-beam system or a double-knock accessory is required. The latter is more economical, although it means a slight reduction in security.

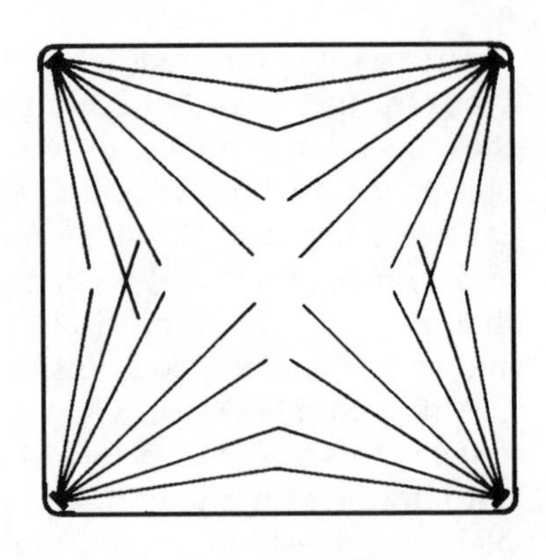

Figure 34 *A large area can be covered with four PIRs in the corners.*

Another problem is the effect of weather, particularly condensation, ice and snow. Built-in heaters keep the projector and receiver clear, but not the reflectors. Microwaves are not affected by these, and are thus the best method of protection where adverse weather conditions are likely.

Passive infrared detectors can be used for outdoor protection, but the problem is that of stray animals. Sensors having detection zones not tilted but directed parallel to the ground considerably reduce the possibility of false alarms from this cause, providing the animal does not jump up into the plane of detection. A double-knock or pulse-counting facility should further reduce the risk. The floodlight type has much to commend it here, as false alarms are of little consequence. Even if PIR detectors connected to the alarm system are used outside, it is well worth while augmenting them with PIR floodlights.

The range of a single detector may be insufficient to cover a large outside yard, but a combination, such as one in each corner, could protect up to 100 ft (30 m) long sides of a square and most of the internal area (Figure 34). For distances longer than this, beam microwave or active infrared systems should be employed. For perimeter fence protection, though, inertia sensors are the most effective.

7 Sounding the alarm

It is a thought-provoking fact that all the technology, planning and work that have gone into the design and installation of a security system, with its sensors and sophisticated control circuitry, are in most cases all to the end of just ringing a bell. Of course, it is the effect of the bell that is the important thing, but it does emphasize the importance of the sounding device. If it is ineffective or can be neutralized, the whole of the rest of the system is useless.

As stated earlier, it is prudent to have at least two bells if not three. These can be situated at the front and back of the building and also inside. Even if the outside bells were ignored or incapacitated, few intruders would have the nerve to enter premises in which a loud alarm bell was sounding.

The important thing with any alarm sounding device is that it should be loud and also strident. Not all bells score well on both points. Some have quite a modest sound output compared with others, while many sound quite melodious and not at all as nerve jangling as they should.

Sound levels and loudness

To understand the volume ratings of different sounders we need to know something of how sound is measured. The much misunderstood decibel (dB) is the unit used, but it does not express an absolute value such as an inch, a pint, a metre or a litre. The reason for this lies in the way our ears perceive sound. Loudness follows a logarithmic rather than a linear scale. So a sound that measures twice the level of another doesn't seem twice as loud to our ears, and one that sounds twice as loud would actually be many times the level of the other.

So to specify the apparent loudness of a sound, we have to use units that compare the measured sound with some standard reference level, and then progress from there in a logarithmic manner. The reference level is the lowest hearing threshold in a healthy young adult, which is a sound pressure of 20 microPascals (μPa). A sound having a pressure of 20 μPa thus has a ratio of 0 to the standard and so is given the value of 0 dB. All other sound levels are also expressed as a ratio to the standard.

Some common decibel ratios for sound pressures are:

6 dB	× 2
10 dB	× 3
12 dB	× 4
18 dB	× 8
20 dB	× 10

The ear does not respond equally to sounds of every pitch (frequency), so for example, a sound pressure that measures 70 dB at a mid-frequency will sound a lot louder than the same measured level at a high frequency, or a low one. Our ears are more sensitive to sounds in the middle of the range. To compensate for this, what is known as a weighting curve is applied to all measurements. There are five different curves which are described by the letters A–D and S; these being designed for different types of sound such as speech and aircraft noise, also for different sound levels because our hearing curves differ between loud and quiet sounds.

In practice most of these have been dropped for various reasons leaving just the A curve. Sound pressure levels measured according to this curve are thus denoted by the term 'dBA'. However, as this is the only curve in general use the 'A' is commonly dropped leaving just 'dB'. In official and technical descriptions though, the full expression 'dBA' will be encountered to avoid confusion with other non-audio decibel measurements.

If all that seems a bit too technical, we can get some idea of the levels of everyday sounds by the following values:

Inside room in quiet neighbourhood	20 dB
A watch ticking	30 dB
Spoken whisper at 1 m	45 dB
Good speaking voice at 1 m	60 dB
Vacuum cleaner at 1 m	75 dB
Average disco level	100 dB
Pneumatic drill at 1 m	105 dB

Anything over 80 dB is usually considered loud, while 100 dB is unpleasantly so. The threshold of pain is reached at 130 dB, and exposure to this level can cause permanent hearing damage in a matter of minutes.

The decibel is used to specify a difference in level between two sound sources such as two bells, or those at different distances from the same source. As everyone knows, sound levels decrease with distance, but this may be overlooked when considering the siting of a bell and also comparing specifications. The levels of some sounders are specified at 3 m, while others are specified at 1 m. This makes the latter appear on paper to be louder than they really are, when compared with those quoted at 3 m, but that is the specification now most commonly used.

Sound pressure for most sources drops in proportion to the distance, so at twice the distance the pressure is a half, at three times the distance it is

Table 2 *Distance at which 70, 65 and 60 dB are obtained for a given sound level at 1 m*

dB at 1 m	*70 dB*		*65 dB*		*60 dB*	
	ft	*m*	*ft*	*m*	*ft*	*m*
83	14.6	4.4	26	8	47	14
86	21	6.3	37	11	67	20
90	33	10	60	18	107	32
93	47	14	84	25	150	45
96	67	20	120	36	214	64
100	107	32	186	56	333	100
103	150	45	266	80	480	144
106	210	64	380	114	668	203
110	340	102	600	181	1073	322
113	480	144	853	256	1520	456
116	680	204	120	362	2150	645
120	1080	324	1916	575	3413	1024

a third, and so on. The decibel value for a ratio of 1:3 is 10 dB, as given on page 84. So a bell rated at 90 dB for 1 m is the same as one specified at 80 dB for 3 m. This point should be carefully noted when comparing specifications.

While the bell needs to be mounted high enough to be out of reach of tamperers, it should not be so high that the sound level is excessively attenuated. At a height of 6 m the sound at ground level is only a sixth of its rating at 1 m, which is 16 dB less.

This attenuation with distance underlines the point made about having a sounder at the rear as well as the front of the premises. The distance from the front to the back of a house along the shortest path could be 20 m or more, and the sound from a bell at the front would thereby drop to a twentieth, or 26 dB less. The minimum sound level at which a sounder can be deemed audible above moderate ambient noise is 60 dB. So if the bell is not rated at more than 86 dB at 1 m, it may not be heard at all at the back, and so will lose its deterrent force.

The sound level may be rather more than these figures suggest, owing to reflections from the walls of the building. However, if the premises are detached and remote from others or from a frequently used thoroughfare, the bell may not be heard by anyone who could raise the alarm. Intruders may take a chance on that and ignore the sounding bell. In such a situation other warning devices are available, and are described later.

Table 2 shows the theoretical distance at which 60 dB will be obtained for a given level at 1 m. This is in still air; any breeze will make a considerable difference, increasing the range in the direction to which it is blowing and decreasing it in all others. The normal temperature gradient, whereby air gets cooler with height, causes sound to refract upwards

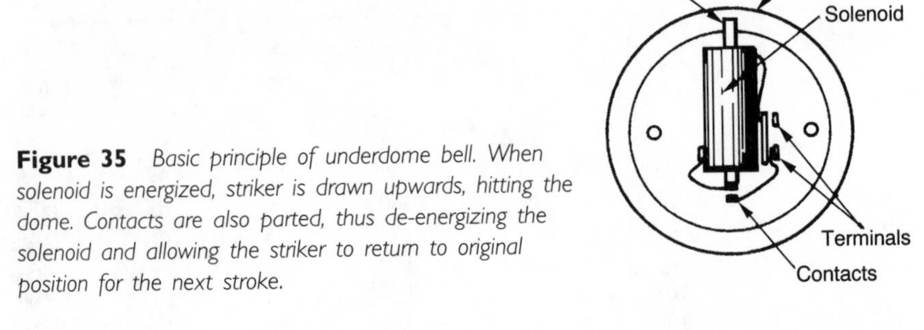

Figure 35 *Basic principle of underdome bell. When solenoid is energized, striker is drawn upwards, hitting the dome. Contacts are also parted, thus de-energizing the solenoid and allowing the striker to return to original position for the next stroke.*

and so be reduced at ground level. This reduces the range. Sometimes there is a temperature inversion, when air above the ground is warmer than at ground level. This causes downward bending of the sound waves and increases the range at which sounds can be heard. The table is thus a rough guide, as much depends on local atmospheric conditions.

Bells

The most common alarm sounding device is the bell. The operating principle is quite simple. A current is applied to a coil, termed a solenoid, and a plunger is magnetically attracted through its centre to strike the rim of a gong (see Figure 35). As it does so a pair of contacts is separated and the current through the coil is interrupted. The plunger is returned to its rest position by means of a spring and the contacts close, thereby reapplying the current and so causing a further strike.

The distance travelled by the plunger determines the quality of the ring. If it is too close to the gong, one stroke is dampened by the striker returning too soon for the next, resulting in a clattering sound rather than a ring. If the plunger is too far away, it does not hit the dome with the full force and the strokes come at too slow a rate, so producing a weak sound.

There is thus an optimum distance, which can be set by a screw adjustment of the contacts or by rotating the slightly eccentric gong. The older type of bell had the striker mounted on an arm, with the coil and contacts located at the side of the gong. Modern bells, especially those for alarm work, have everything concealed underneath the gong.

This type of bell with interrupter contacts will work off AC or DC, because polarity does not affect the action. It can thus be powered by a mains transformer of appropriate voltage, by batteries or, as is usually the case, by the DC output from a control unit. Bells intended solely for AC operation do not have interrupter contacts, and some are designed to work directly from the mains supply, although these are not used with domestic intruder alarm systems.

The control box supply is usually 12 V DC. Other bell ratings are 6 V, 24 V and 48 V, the higher ones being mostly used for fire detection systems. These voltages are not critical and many bells are rated over a range such as 6–12 V or 12–24 V. With these, two current and sound pressure ratings are given, one for each voltage extreme.

The sound output is greater at the higher voltage within the stated range, as is the current. Current rating is important where several bells are to be used on the same system, as the total must not exceed that specified for the control box. Also important is whether the standby battery could sustain an alarm with all bells sounding for the full alarm period (usually 20 minutes), and then be ready for a further alarm without replacement or recharging.

A bell should thus be chosen that has a moderate current rating but high sound output. A good bell should take less than 0.1 A (100 mA) at 12 V, and deliver in the region of 96 dB at 1 m. Not all models achieve this standard, either taking more current, delivering less sound, or in some cases both.

The design of the gong has much to do with the sound level and also the tone. Those having a dished effect in the centre appear to give a harsher and louder sound than the conventional dome shape, as do also those made of a heavier gauge steel. Most alarm bells have a gong that is 6 in (150 mm) in diameter, but some are also available with 8 in (200 mm) and, more rarely, 10 in (250 mm) gongs. A 200 mm bell usually has the same coil and striking system as the smaller version of the same make but gives about 3 dB greater sound volume, which is over 40 per cent more. This does not necessarily follow when comparing different makes, as the larger gong may be of thinner gauge steel and hence of smaller mass, and could thus actually produce less sound than the smaller one.

Bells are available from many manufacturers, but one that is especially recommended is the B6D12 and the B8D12 made by Tann Synchronome of Clevedon, Avon. These are 150 mm and 200 mm units respectively, the 200 mm giving 5 dB more sound than the 150 mm. They have dished non-removable domes, and a high output that makes most of the competitors sound like door bells. They also have a low current rating of 95 mA, so several could be run from the normal control unit without exceeding its current output. The units are waterproof and can be mounted without a case. A waterproof gasket can be supplied by the makers for open mounting. It is the bell specified for the Sureguard system described in Chapter 14.

Another type, known as the *centrifugal* bell, uses a small motor to rotate the striker at high speed, repeatedly striking the inside rim of the gong.

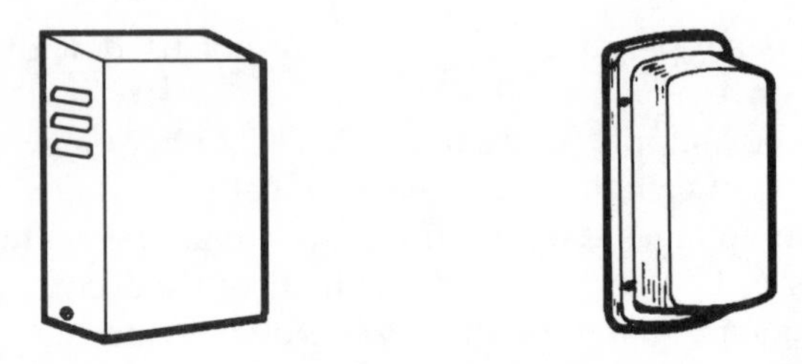

Figure 36 *Bell boxes: left, steel bell box with side louvres; right, totally enclosed fibreglass bell box.*

The current is usually higher than for the best solenoid types, although it is about the same as most of the high-current ones. The sound output is about the same. The construction is less robust than the solenoid bell with more moving parts, so there seems little to recommend it over the solenoid type.

Enclosures

It is the common practice to enclose the bell in a case. Materials used are polypropylene, which has moderate strength but does not rust; polycarbonate, which is also rustproof but much stronger; plastic coated steel for maximum strength; or stainless steel. Enclosing has two objects: to protect the bell from the weather and also from tampering (Figure 36). To increase security, cases are usually fitted with anti-tamper microswitches under the cover.

Inevitably, enclosing the bell reduces the radiated sound. This is generally considered to be the price to pay for the security and weatherproofing a box affords. However, vented boxes can actually be a security hazard as a bell can quickly be stifled by filling the box with sealing foam squirted through the vents. Boxes are not as indispensable as is often thought; a far greater protection is to mount the bell in a position that is inaccessible without a ladder.

If access is obtained, many bells could indeed be easily neutralized by removing the gong. However, the Tann Synchronome models have a gong centre screw that can only be removed with a special tool, and are virtually impossible to silence by jamming or physical attack (Figure 37). These are also weatherproof and are well worth considering instead of the more usual enclosed bell. A further advantage is that it is evident that an alarm system is indeed installed. Dummy bell boxes are frequently fitted as a deterrent, but the more knowledgeable of the criminal fraternity are getting wise to this. An open bell leaves no doubt.

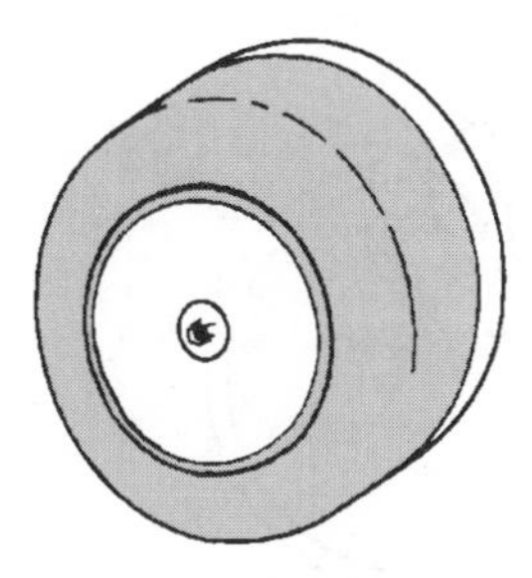

Figure 37 *Tann Synchronome bell.*

Self-activating bells

The wiring to the bell may be vulnerable to tampering in some high-security situations. It can be protected to some degree by means of anti-tamper loop wiring as for the sensors, even if an enclosure with an anti-tamper micro-switch is not used. However, as the bell has such a vital role to play, special measures are needed to combat possible attacks on the bell wiring.

One method of doing this is to use a self-activated bell. The bell is housed in a box with its own battery and a hold-off circuit. This is effectively held off by a voltage from the control unit, so if the wire is broken or short-circuited, the hold is released and the bell is sounded.

Actually, more than just the wiring is protected by modern units. If the control box is destroyed, or the power supply is cut off, or the system standby battery runs down, the effect is the same: a releasing of the hold-off and sounding of the bell. The unit thus affords a back-up to the whole alarm system.

At one time, dry batteries were housed in the bell box and were the weak point of the system. Unused batteries deteriorated and needed replacement, but were not easy to test or replace. With modern units, rechargeable Ni-Cd batteries are used, and these are automatically kept charged from the control unit. A further feature is that a normal alarm, that is one not involving operation of the self-activating circuitry, does not use current from the battery but draws it from the control box in the usual way. The battery is thus kept fully charged for any occasion when it may be actually needed.

Self-activating modules are available which can be used in conjunction with most bells (Figure 38). Most, though not all, come complete with the battery fitted to the board. The modules are usually housed in the box with the bell, and an anti-tamper microswitch is usually included (Figure 39). It is possible to use a module with an open bell, by housing it in a locked case on the inner side of the wall on which the bell is mounted. Wiring from the module then passes straight through the wall to the bell.

When self-activated, the bell is stopped by reapplying the hold-off voltage. If this ceased because of damage to the wiring or the power supply

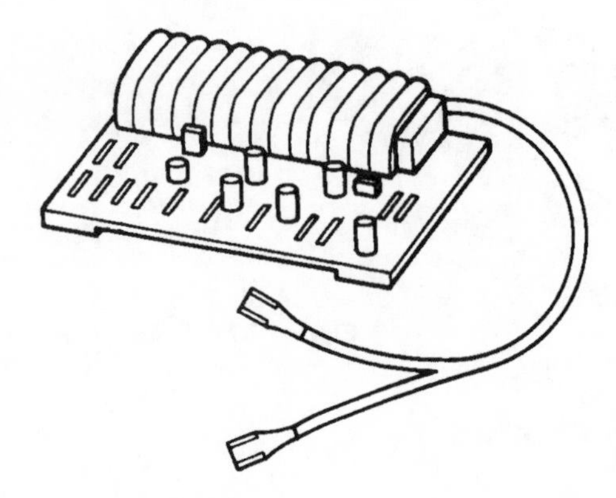

Figure 38 *Self-actuating module for fitting to standard bell.*

Figure 39 *Self-latching bell in box. Latching module and terminal strip can be seen; space is for housing internal batteries.*

it may take some time to restore. However, the on-board battery capacity is limited to less than an hour of sounding, depending on the bell current, so there is no fear of the bell continuing to ring for hours until someone can deal with it.

Although security is increased by self-activation, it is not totally defeat-proof. As hold-off is obtained by a voltage supplied from the control unit, it can also be achieved by connecting an external battery across the hold-off circuit. This can be done by means of a battery connected to the appropriate wires via pins pushed through the insulation. The alarm circuit can thus be neutralized while the wires back to the control box are cut. If left so connected, the battery can hold off the alarm for several hours.

To succeed, the correct pair of wires out of the four usually run to the bell would have to be intercepted, and the polarity of the battery would have to be correct. This could be discovered by pinning the four wires and checking with a voltmeter to determine which pair carried the hold-off voltage, and its polarity. This is not the sort of thing the casual amateur thief would be likely to do, but it is not past the ingenuity of a professional with a high prize in view.

The best protection for bell wiring is to conceal it where it cannot be reached, in conduit, under flooring, and buried in plaster; it should always be brought out directly behind the bell and never run along part of the wall exterior. If this is done, self-activation is unnecessary for most

Figure 40 *Sirens: left, high-power horizontal siren; right, smaller vertically mounted siren.*

alarm systems, certainly for domestic ones. Where high security is needed, self-activation can be used, but the bell wiring should also be physically well protected and hidden.

Timing modules

As pointed out in an earlier chapter, it is desirable and in many places mandatory to limit the bell sounding to around 20 minutes. The intruders should have disappeared by then, either under their own steam or in the back of a police car. Further sounding only causes annoyance to those nearby, especially at night and if, as so often, it is due to a false alarm.

All modern control panels have integral timers which so limit the bell sounding, but many older ones do not. If such an untimed system is already installed and is satisfactory in all other respects, there is little point in changing it. A timing module can in most cases be fitted to exercise the necessary control. Timing is usually adjustable up to around 45 minutes maximum. In some more remote areas, a longer sounding than 20 minutes may be desirable to ensure that attention is attracted.

Sirens

Where warnings are required to be extra loud or to sound over much longer distances than usual, a siren is often used. The mechanical type consists of an electric motor driving an impeller which forces air through vents in the casing in such a way as to produce a loud tone, usually of a raucous nature to arrest attention. The complete unit is totally enclosed and mostly is designed for outdoor applications (Figure 40), although smaller non-weatherproof versions are available for indoor use.

Decibel ratings usually range from 100 to 120 dB at 1 m, which is louder than a pneumatic drill. Makers commonly quote distances at which the device is audible. These usually assume a minimum audibility requirement of 60 dB and so generally correspond with Table 2. As already mentioned, these distances depend on local conditions. It will be noticed

Figure 41 *Electronic sirens.*

from the table that there is a considerable difference in distance for just a few decibels, especially at the high end. For remote locations then, a few decibels extra can make all the difference as to whether the alarm is heard or not.

The high-power models are usually driven by mains voltage, and so most control panels require a relay to run them. Some panels have a switched mains outlet to which these sirens can be directly connected. It should be noted, though, that should the mains fail, and the system switch itself over to the standby battery, no alarm can be sounded if the sole sounder is mains operated. A bell or 12 V siren should also be installed to sound in the event of an alarm during a mains failure.

Sirens with 12 V motors can operate directly from most control panels which deliver 1 A to the sounder, but some caution is needed. Average siren current is usually just under 1 A, often 0.9 A which is within the control panel rating, but the current at the instant of starting can be a couple of amperes or more. It is as well to check with the control panel makers if the panel will take this momentary excess load.

If such a siren is used no other sounder can be connected as the total output current available is thereby taken. If others are required, a relay will be necessary that will switch the total load to be applied.

An alternative is to use a solid-state electronic siren (Figure 41). These consist of an oscillator, an amplifier and a horn-type loudspeaker, all contained in the same unit. Output is slightly less than for the loudest motorized sirens, but the current is much less. Some models take as little as 0.02 A (20 mA) yet produce up to 107 dB at 1 m. Others which deliver some 116 dB take 0.35 A (350 mA). Although the latter are comparable in sound output to the motorized sirens, the current is little more than a third. They have no starting current problems, and other additional sounders can be comfortably accommodated within the 1 A maximum output of the control panel.

A horn speaker is used because this produces more volume per watt than any other type. To take the minimum space, the horn is re-entrant, that is folded within itself. However, it is a highly directional device and so needs to be pointed towards the area from where attention is be attracted.

The frequency or pitch of the generated note varies between models. The ear is most sensitive to frequencies between 2 and 4 kHz; however, tones in this region and above are attenuated with increasing severity as distances extend beyond 50 m. So at long distances the losses cancel any advantage gained by using frequencies in this band. Those just below it, from about 800 Hz to 1 kHz, are still very effective, yet they carry considerably further. They are therefore the most commonly used.

Electronic sirens often offer a choice of effects. The most common are: a fixed steady tone; a warbling effect between two tones; or a single tone that is pulsed about twice a second. These can be selected at will by using the appropriate connections, and they offer the possibility of having different sounds to indicate different things. One could be used to warn of intruders, another for fire and, in the case of a factory, a third (perhaps the steady note) could be used as a finishing signal.

Another type of electronic sounder is the piezoelectric siren, which uses a piezoelectric transducer instead of the horn. These also give a high output with low current consumption. Their main disadvantage is that the tone is high-pitched which, although it attracts attention, does not carry very well outdoors. They can be used effectively for short-range outdoor use, or as an internal sounder.

The electronic siren is thus versatile and easily matched to any control panel. It is of particular value in industrial situations when large areas need to be covered, or for homes which are remote from any habitation and from which an alarm sounding would have to cover some distance to be heard. For houses in urban areas, though, the bell is to be preferred. In city streets, a siren can be confused with vehicle alarms and with police car, ambulance and fire tender sirens. When an insistent bell is heard, there is little doubt that an intruder alarm has been set off in nearby premises.

Illuminators

Although not strictly *sounding* the alarm, some form of illumination or visual indication linked to the alarm system is often desirable. Consider a common situation where there is a row of houses each having an alarm bell outside. Suddenly one starts to sound – but which one? It is frequently difficult to determine, and often requires the listener to stand under each one in turn to identify it. A visual indication will quickly show which premises are being attacked and so help can be summoned with minimum delay.

This can be done by means of a lamp mounted on or near the bell box. Some boxes are translucent, enabling the lamp to be mounted inside if room permits. The light source commonly used is the xenon tube, a tube filled with xenon gas which emits a brilliant flash when a high voltage is momentarily discharged through it. The high voltage is produced by the integral associated circuit in the beacon, but the actual supply voltage is usually 12 V, although there are also 24 V and mains voltage versions. The light is commonly though incorrectly called a strobe.

The brightness is determined by the electrical energy discharged through the device and is rated in joules. A typical value is 5 joules, but there are higher-powered ones up to 12 joules. The joule is a power of 1 watt sustained for 1 second, but as the flashes last for only a fraction of a second, the light intensity is equivalent to that of a high-power filament lamp. Even the 5 joule devices can be visible for several miles, depending on background ambient illumination.

Normally, the flash rate is about one per second, but it can be set to a higher rate of up to two per second if required. The life of the xenon tube is some 5 million discharges or more, so at the highest rate of 120 per minute, and for a period of 20 minutes operation for each alarm, it should last for at least 2000 alarms. It is hoped that no one would be so unfortunate as to ever need a new tube!

The current taken by the 5 joule lamp is in the region of 0.1 A (100 mA), and so the unit can in most cases be connected directly across the bell without exceeding the control panel output current. The larger ones take 2 A or more at 12 V and so require an output relay to operate from a normal system. If the control panel has a mains switching facility, a mains-voltage xenon could be used directly if high power is required, without the need for an extra relay.

As well as giving visual warning and identifying the source of an alarm, lamps can be used to provide steady illumination or to terrify and confuse the intruders. The control panel mains switching facility or a separate relay can be connected to high-power filament lamps to floodlight a yard or the area around the premises. Several lamps can be used providing the total power does not exceed that specified for the relay or panel. These will remain on for as long as the alarm is activated.

Another interesting device is the flasher module, which is connected to the alarm output circuit. This consists of a relay that operates a changeover switch once per second. Two high-power (up to 500 watts) lamps can be connected so that they go off and on alternately: one comes on when the other goes off. If the lamps are situated well apart, the alternate illumination from different angles gives the impression of being surrounded by flashing light and will wreck the nerves of the boldest intruder, especially as the alarm will be sounding at the same time. All these devices are linked to the alarm system and so come on when the alarm is triggered.

The PIR-controlled lamps described in Chapter 6, which operate independently, are also well worth considering. These are activated by body heat over a range of several metres, and switch on the associated floodlight whenever they detect movement of a warm body within their range. They can thus be used within the protected perimeter as a back-up system, or at the perimeter to discourage attempts at entry. Once activated, the light remains on for a pre-set period of from seconds to minutes.

A further type of security illuminator is the rotating mirror beacon such as used on police cars and ambulances. These are available in different colours, though not usually blue to prevent misrepresentation. The lamp is rated at 48 watts, but the parabolic reflecting mirror concentrates the light into a powerful rotating beam. This is most effective when there is space all around rather than when the beacon is fitted to a wall.

Remote signalling

Our concern so far has been for the alarm system to actuate an on-the-spot sounder. Its purpose is to scare off the intruders before valuables can be stolen or damage committed, and to summon the police who have been alerted by a neighbour or passer-by. The weak link is that it could be some minutes before the alarm is noticed and the police informed, and further minutes before they arrive. Where there are valuable articles, daring intruders may take a chance on this and stay just a few minutes to grab as much as they can before disappearing.

So the sooner the police can be informed the better. At one time it was possible to have a private telephone wire to the local police station, and a signalling system would send an alarm on being activated; faulty or broken telephone wires were also indicated, making the system virtually foolproof. Unfortunately, the high incidence of false alarms has caused this facility to be withdrawn. It is still possible to summon help using automatic telephoning by either telediallers or digital communicators, though this is less effective than was the direct-wire system.

Telediallers can be analogue or digital. With the analogue ones, a telephone number is recorded on an endless tape loop followed by a verbal message. When activated by the alarm system, the tape starts and the number is dialled, then the message is delivered. The system is not unlike a telephone answering machine except that it does the dialling instead of waiting for a call. Several different numbers and messages can be sent in succession, so a 999 call can be followed by calls to friends or neighbours.

One advantage over the direct-wire system is that a special line is not required: the normal telephone line can be switched over to the dialler when the householder is are absent. There are two snags with these devices. Firstly, unlike the answering machine, you cannot record your own message; it must be done by the company that supplies the equipment. This is because the telephone number to be dialled has to be specially recorded using pulses before each message to which the telephone exchange will respond. If there are any telephone number changes, the whole thing must be re-recorded.

The other drawback is that the machine dials each number once only. If it does not get through first time it doesn't try again. However, if several numbers are dialled the chances are that at least one will be contacted. It should be noted that in some areas the police will not even accept calls from telediallers owing to the high rate of false alarms. It is indeed unfortunate that what could be a valuable security aid is made unavailable by others' carelessness in the installation or operation of alarm systems. It underscores the high priority which should be given to preventing false alarms.

Some telediallers are dual purpose. A separate tape can be actuated by fire detection devices and a 999 call sent to the fire services as well as additional calls to other numbers.

The second type of teledialler is the digital dialler. This also makes use of a pre-recorded tape to send a suitable message, but the 999 number is contained in a digital memory instead of on tape. In this it is like those telephones that have a memory for storing frequently used numbers. The 999 call is the only number it will dial, but the message, of some half a minute duration, will be continuously repeated for up to four minutes. This overcomes the possibility with the analogue dialler that part of the message may be lost owing to noise on the line, momentary distraction of the person taking the message, or other cause.

An alternative to both analogue and digital diallers is the digital communicator. These do not use tape at all and so are more reliable; they are also cheaper. When activated, they dial a computer at a central receiving station which sends back a 'handshake' signal. If this signal is not received by the communicator it assumes it has not got through and so keeps dialling until it does receive it. This is an improvement on the diallers, which give up after the first attempt because they have no means of telling whether contact has been made.

When the return signal is received, the communicator sends a digital code which identifies the sender, then transmits a coded message. The computer finally sends back an acknowledgement signal. If this is not received, the communicator redials and repeats the process. Most communicators can send only one type of message, but others can send different codes to correspond to different situations: an intrusion, personal attack (actuated by a panic button), or fire. Some can send up to eight different codes. Further facilities offered by some models are the dialling of two telephone lines to the same control station to ensure a connection, and the dialling of auxiliary numbers in addition to that of the receiving station.

On receiving an alarm signal, the station contacts the police and any other party designated by the subscriber. Even with these, there is some reluctance on the part of the police to respond as a result of so many false alarms. They put the onus on the station to filter them out, but the station has no means of identifying a false alarm from the genuine ones. So their response is to refuse to accept subscribers whose alarm system has not been installed by an approved installer. This unfortunately cuts out the DIY installation.

Receiving stations are run and maintained by private security firms and also British Telecom; most areas have at least one. These are generally advertised in the *Yellow Pages* as a 24-hour central station service. Payment is usually a fixed annual subscription plus a charge each time the service is used.

Diallers and communicators often have the facility of controlling the sounders that are operated by the alarm system. In particular, this means delaying the sounding of the bell for a set period to enable the police to arrive and catch the culprits. If a fault is detected on the telephone line they switch to instant sounding. As pointed out earlier, delayed sounding is a questionable practice as thieves could make a valuable haul or cause much damage and be gone before the police turn up.

Modern digital communicators are quite small compared with the tape dialler. Many are just a printed circuit board that can be installed inside a control panel, and some control panels already have them fitted.

The details to be transmitted are contained in an electronic memory device called an erasable programmable read-only memory (EPROM). What this mouthful really means is that it is a memory chip from which data is read but to which data cannot normally be added by the user. The data are placed in it (it is programmed) by a special device which can also erase and reprogram it if required. This is done by the makers or the suppliers.

Radio pager

Another option worth considering by those frequently absent from home, but within a few miles, is the radio pager. This is a small transmitter that is triggered from the alarm system, or from an independent PIR detector in the grounds of the protected premises, thereby giving warning of prowlers.

One example of this type of device operates from a 12 V supply taking 0.05 A (50 mA) when on standby but 1.6 A when active. This is more than the usual control panel output, so an extra relay is required. The pager is supplied complete with aerial, and will transmit up to four miles. Remember this is four miles 'as the crow flies' and is further than a four mile journey by road. However, hilly terrain or nearby high-rise buildings may reduce the range.

The receiver, which has a pocket clip, is a breast pocket sized unit that flashes a light and sounds a bleeper when activated. It is powered by two small alkaline batteries, but rechargeable Ni-Cd units could be fitted.

Any number of receivers can be used, but they must all be tuned to the same transmitter frequency. A number of different frequencies are employed to reduce the possibility of one transmitter activating other receivers in the same area.

The device thus informs the householder of intrusion without the expense and problems associated with telephone links, and when he/she is mobile. The disadvantage is that at $\frac{3}{4}$ in (19 mm) thick they are just a shade bulky to carry during an evening out in a dress suit breast pocket.

8 Planning the system

As pointed out previously, security is relative. It is virtually impossible to make any premises absolutely invulnerable; bank vaults are probably the nearest thing to it, yet even they are broken into at times. The important thing is to achieve adequate security sufficient for the degree of risk.

For example, the average home is usually the target of the opportunist thief, and if it presents too many problems to break into or obviously has a good alarm system, he will go elsewhere where less trouble and risk are involved. Even so, it is wise not to underestimate him. He has probably learned many tricks from others on the street or in prison, and knows how to spot any weak points that a householder may overlook. So it is better be a little over-secure than under. However, extremes can be self-defeating as they are not usually necessary, can make life a misery, and as a result are often relaxed after a while, making the premises more vulnerable than they would have been with more moderate protection.

Homes that house special valuables such as art treasures, cups and trophies, and expensive jewellery, or those that appear affluent, especially if occupied by someone living alone, can be a special target of the more experienced and determined crook. With these, a higher degree of security is required, both physical and electronic.

Another phenomenon which sadly is increasing is that of thugs breaking into quite modest homes while the occupant is present, and using extreme violence, often to no apparent purpose. This is something which at one time was unheard of or extremely rare, but now must be taken into consideration when planning security measures.

Apart from these factors, there are others that also determine the degree of security required. The presence of a large council housing estate in the vicinity, a nearby public house, school, night-club or sports facility, can all increase the risk. Premises in dimly lit side roads or those with no nearby houses or passing traffic at night are also more vulnerable. Those at the end of a terrace are at greater risk than premises within it, as are detached or semi-detached buildings. Lock-up shops are a more attractive target than those with occupied living accommodation. Also to be considered is the crime record for the neighbourhood. Some, such as inner city areas, have a high incidence of crime, while for others it is comparatively low.

So it is a matter of exercising good judgement. A good standard of general security with a reliable alarm system should be considered a minimum requirement, but adding the higher-security features that have been noted in the foregoing chapters depends on the risk considered to be present.

Become a burglar!

Well, in thought if not in deed! Try to look at your premises through a burglar's eyes, put yourself in his shoes, take a walk around from the outside to see where you would break in if you had to. If the cap fits uneasily, imagine instead that you have locked yourself out and you have *got* to break in somehow; but to save embarrassment you want to do it unobserved by passers-by or neighbours.

One possible difference between you and the burglar, though, is that he is likely to be in his teens, to be agile and able to climb about, and to have a good head for heights. This means that you should not overlook possible first-floor entries, especially if there is a nearby wall, post, pipe, tree or any other means of getting up there (Figure 42). The burglar will probably be a lot slimmer than you are, so narrow windows or fanlights will be no problem to him. Some victims have expressed surprise that entries have been made through unbelievably small apertures: in one case it was just 8 in (200 mm).

Remember the point made earlier that the intruder does not want to leave by the same way; his aim is to open up an exit route through the front or back door, preferably the latter as he can then escape if he hears someone returning through the front. The importance of making all exit doors unopenable without a key can thus be appreciated. The intruder then has to face leaving the same way that he entered, so risking being seen – and this exit could be tricky if he had to take it in a hurry. Many a burglar has had a nasty fall and been badly injured doing just that.

The desire of the intruder to be unobserved means that points affording some cover are especially vulnerable. Trees, bushes, walls and advertising hoardings are all very welcome to him. For the same reason, poorly lit areas away from street lights are his favourites too.

Detached, semi-detached, or end-of-terrace premises are also preferred because there are neighbours on only one side, and so there is less possibility of being observed or heard. In addition, there is likely to be a freer escape route from a rear which is not hemmed in by two neighbouring gardens.

After taking a criminal's eye view of your premises, note down the possible entry points in order of vulnerability. Ground-floor rear windows will undoubtedly head the list, followed by other ground-floor

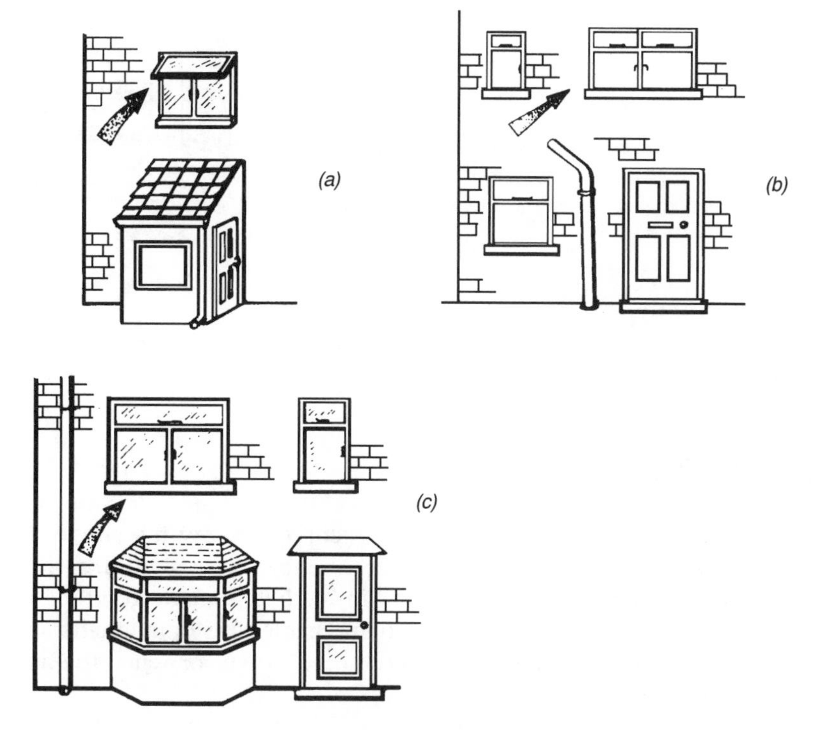

Figure 42 *Upstairs windows are vulnerable where there is easy access from: (a) porch or lean-to; (b) drainpipe; (c) bay window; or any other nearby support. These should be wired, and kept closed when no one is at home.*

windows and back or side doors. If there are any outhouses at the rear through which access can be obtained to the main building, these are likely possibilities. Then comes the first-floor windows that can be climbed up to. If you have a cottage-type dwelling with low ceilings, the first-floor windows will be lower than normal, and possibly accessible by a burglar standing on the shoulders of an accomplice. In this case all first-floor windows are vulnerable. Flat roofs too are easily broken through.

Do not omit any reasonable possibility; remember the old adage that the strength of any chain is in its weakest link. Leave just one point unprotected, and that will be the one through which the intruder will get in. Then all the work and expense on the rest will have been wasted. There have been many sad cases where break-ins have occurred and losses suffered in homes that had seemingly been well protected with an alarm system and security devices, but in which just one point had been overlooked or neglected.

Before even thinking about an alarm system, get the physical security

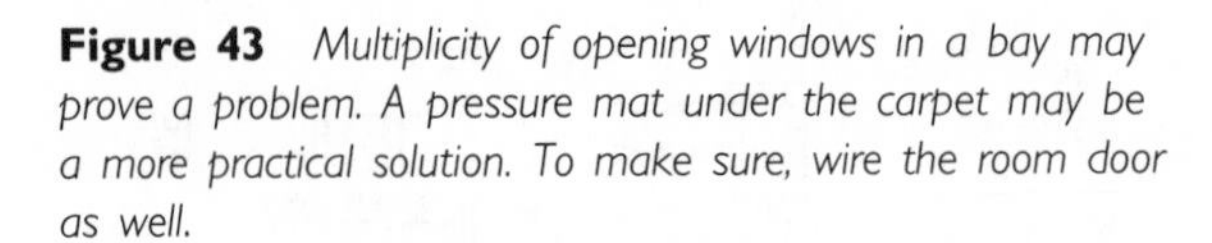

Figure 43 *Multiplicity of opening windows in a bay may prove a problem. A pressure mat under the carpet may be a more practical solution. To make sure, wire the room door as well.*

in good shape. That is the first line of defence and the first priority. We will take a look at some specific points.

Make an inventory

First, get those ground-floor windows secured. Follow the suggestions made in Chapter 1. Locking catches or lockable bolts should be fitted, and puttywork carefully examined and immediately made good if any is cracking or coming away. The position and cable runs of magnetic alarm sensors can be planned and tentatively noted down for when the alarm installation is made.

Any windows that seem especially at risk can be measured up for a decorative metal grille. Or, if the situation permits, the site beneath can be earmarked for a rosebush, to be planted as soon as the suitable season arrives. Non-opening windows can be fitted with window foil for connection into the loop. If there are no other sensors in that area of the house, foil can be fitted anyway, and left unconnected. Its appearance will serve as a deterrent.

The back door is the next focus of attention. Lockable bolts (rack bolts) should be fitted top and bottom, and a double-locking deadlock. Does the door open outwards? If so, dog (hinge) bolts are a must for the hinge side of the door. How sturdy is the door? Could it be burst open with a hefty shoulder charge or a kick? (No, don't actually try it!) Does it contain glass? Is the frame reasonably substantial? Replacement may be wise and cheaper in the long run if any of these seem deficient.

Now we can look at the front. The windows should be checked for physical soundness just as those of the rear, and the position of sensors determined. Bays having a number of opening windows may pose a problem in having to wire each one for an individual sensor. Is it necessary for all of them to open? Could some be screwed up, leaving just one or two to open? Alternatively, or in addition, a pressure mat could be installed inside (Figure 43). If though the bay is secluded and out of sight of the road, or high security is required, it would be wise not to rely just

Figure 44 *Glass-panelled front doors are extremely vulnerable, especially if ordinary locks are fitted. Such doors should always be wired to the alarm. Solid wooden doors with security locks need not be wired as these are not likely to be attacked. All back doors into the main building should be wired.*

on a pressure mat. As at the rear, non-opening windows should be fitted with window foil.

Glass panels in the front door make it extremely vulnerable as access can easily be obtained to the lock (Figure 44); also, doors having a lot of glass are structurally weak. A deadlock or deadlatch is absolutely essential in this case, but is strongly recommended with any front door.

Most modern UPVC doors have locks that engage bolts in the top and bottom of the door as well as the sides, and this affords a high level of security. However, these are not totally intruder-proof if they have a glass panel. It is true that when the door is locked, the inside lever is immobilized so that an attempt to open it through broken glass is frustrated. But this is so only if the door *is* locked. When it is unlocked, as is likely when the house is occupied, an entry could be quickly made by this means. Many intruders today do not baulk at breaking in when the occupants are at home, and attacking them. If a new UPVC door is to be fitted and a glass panel is desired, it should be specified that the glass occupies only the upper third of the door so that the door handle cannot be easily reached through it.

A decorative iron grille behind the glass is highly recommended in all cases where there is glass in the front door. It not only prevents entry but deters breaking the glass in an attempt. Furthermore it strengthens the door itself. Last but not least, it looks good too.

Now look at the door frame. Is it substantial? Is there good support at its sides? or is the plaster doubtful? If there is any doubt, get it chipped out and the cavity filled with cement. It can be decorated over without detracting from the appearance.

Ordinary bolts are no protection when you are out as they cannot be bolted when leaving, but they can be bolted by an intruder to keep you out while he escapes. For higher security, two locks well spaced apart are the best option.

How to exit?

A decision must now be made as to the type of exit arrangements which will be most convenient for setting the alarm system. This was discussed in Chapter 3, but here is a brief recap of the options. The choice is between the timed exit from switch-on, where the exit is timed from opening the final door; the exit set, whereby the system is turned on by the final door sensor being actuated; a key-switch or lock-switch which switches on the system when the key is turned in the exit door; a key-switch or lock-switch that just inactivates the exit door sensor; or leaving the exit door without a sensor, so avoiding all the exit problems. Control units differ in the exit facilities they offer, so the choice will be determined by the units available and will govern which of those is chosen.

The timed exit is the usual, but is fraught with possible snags. Especially if there are young children, the family may not all exit before the time expires. When the family is entering with a load of parcels, the timed entry is likely to be even more stressful. For a single person or a couple, the timed exit and entry are not too difficult providing the timers are set to allow sufficient time.

The exit timed from the opening of the exit door is more convenient for a single person leaving, but is worse for getting children through as in the interests of security the time is quite short, just a matter of seconds. A key-switch eliminates all the problems of timing, but necessitates having to carry and use yet another key. In turn this disadvantage is overcome by using a lock-switch. With this the switch is combined with the lock, so that the system is switched on when the door is locked with no need for an extra key. However, not many control units have this remote-setting facility. The lock-switch connected to shunt the exit door sensor can be done with any control unit.

If the front door and its frame are substantial, and the door has no large glass panels and is fitted with high-security locks, the risk of a forced entry through it may be deemed minimal. In this case much inconvenience could be saved, with little loss of security, by simply not fitting the door with a sensor. Then the system can be switched on and off at will, and there will be no exit and entry problems. To be on the safe side, a pressure mat could be fitted in the hall where an intruder would be sure to step. This of course would have to be avoided when leaving or entering, but it would cause little inconvenience as you would soon get used to it. With young children though it would be a different matter, and the mat would have to be dispensed with.

Another advantage of leaving the front door without a sensor is that neighbours can enter when you are on holiday to gather up mail, without having to contend with the alarm system. This aspect will be discussed more fully in Chapter 12.

It may take a while to mull over these points before making the final choice of exit arrangement, so in the meantime attention can be given to the first-floor windows. Take a look again at Figure 42. Are any of your windows like these? If so, they should be fitted with lockable catches or rack bolts and an alarm sensor. Many an entrance has been made through a first-floor bedroom window while the occupants were asleep at night. Protecting these may require some thought as a bedroom window may need to be left open on a hot summer night. Many UPVC windows can be locked in the partly open position. Failing this, some means of restricting the opening should be devised.

Conservatories, kitchens and extensions

Conservatories are very vulnerable: they have much glass, a flimsy door, and a less than substantial roof. It is not really practicable to try to protect these. The important thing is to protect the door that leads from the conservatory into the house. There will be nothing of value in the conservatory; the intruder's aim will be to get through it into the house. The conservatory just affords cover to make him less conspicuous in his efforts to break in.

So, this door must be substantial, well protected physically with security locks and rack bolts, and wired with an alarm sensor. The same applies to many kitchens that are single-storey extensions. Concentrate on protecting the door that leads to the kitchen from the house rather than on the kitchen windows and doors.

The larger living area extension must be carefully considered. Are valuables such as TV and video kept there? If so, the windows and exterior door will need the same treatment as the rest of the ground-floor entrance points. In addition if it is a flat-roofed single-storey structure, there is the possibility of a break-in through the roof. Lacing wire zigzagged across the ceiling and decorated over is one way of protecting it. Another is to fit a vibration detector to one of the main ceiling members.

The area could also be protected with pressure mats or a PIR detector, but these would only operate when the intruder had gained access and done considerable damage to the roof and ceiling. The wire lacing or vibration detector would trigger an alarm at the first attempt at breakthrough, and so minimize the possible damage. Irrespective of the protective measures taken with the extension, it is prudent to also well secure and wire the door from the main building leading into it.

Bungalows

These are at special risk because all the windows and doors are ground-floor ones. In addition, there is the risk of an entry through the roof. Tiles are easily lifted off, roofing felt is torn away and so into the loft. The loft trapdoor is then lifted, to be followed by a short drop to the floor beneath. Exit is then via any door or window that can be easily opened.

The blocking of all exit routes is thus very important with a bungalow, as without such a route the intruder is trapped. Returning the way he came would be very difficult. He may wait for the owner to return, attack him, and thus make his escape. So, if on returning home you see disturbed tiles on the roof, never enter! Phone the police and wait for them before unlocking the front door.

The loft trapdoor should be well secured with strong fastenings underneath that would stand a kick from above. It would be best if it was hinged on one side and bolted with at least two bolts on the other. It should also be fitted with a sensor.

The bell

The final exterior item to plan for is the bell and its location. The main bell should be mounted on the exterior of the building in a position that is inaccessible except by a ladder. This means high up, away from drainpipes, bay windows or other means which could afford access. The bell can be fixed vertically on the wall (most units are fitted in this position) or horizontally under the eaves if space permits. The latter position is easier to install as it means screwing into wood instead of having to drill and plug a brick or stone wall. Also the wiring can be taken easily through the wood into the back of the bell, and the wood acts as a sounding board, so increasing the sound volume (Figure 45c).

Wall mounting means making a hole right through from the inside, and this will govern the position, as it is undesirable for the wiring to come out in the middle of a room wall. It is best for it to pass through the wall between the ceiling and the floor above it so that the cable can run under the floor and drop down through the ceiling immediately above the control box (Figure 45a, b).

The important thing is that the bell should be in a position where it will be both seen and heard. This usually means fitting it on the front of the house which is nearest the road, but not necessarily. The road may be little used, perhaps a cul-de-sac, yet there may be houses nearby facing the side or rear of the premises. As there is little point in putting the bell on the side of a building furthest from those likely to hear it, it would be better to put it on the facing side.

It should also be readily seen when approaching the premises and thus serve as a deterrent. If there is a conflict between these requirements, two external bells should be used, each on a different side of the building. In such cases it is essential that low-consumption units be used,

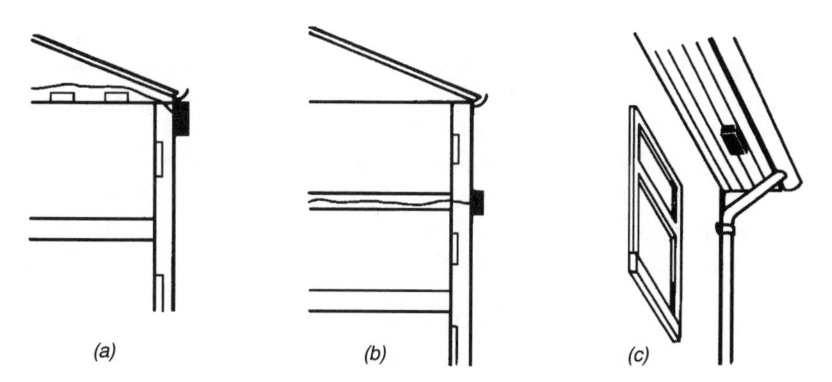

Figure 45 *(a) An outside bell can be fitted above room ceiling level, then wiring can be run above ceiling to be dropped to where the control box is located. If passing through an intervening storey it may be most convenient to take it down a corner or door frame. (b) With three-storey buildings it is best to mount it between first and second floors. This is high enough to be beyond access; any higher would reduce sound volume at ground level. (c) A good place is under the eaves if they are wide enough. Fitting and drilling is easier, and wood can act as a sounding board.*

otherwise the control unit current rating may be exceeded and the life of standby batteries would be curtailed.

Whether or not two external bells are used, it is good practice to fit an internal one. This would rouse sleeping occupants, and it would unnerve intruders who may have braved the external bell to enter. They may even have penetrated the perimeter defences but triggered an internal sensor, in which case the internal bell should scare them out. The internal bell should be mounted where it is difficult to reach and can be clearly heard over the whole house. A good position is at the top of the stairwell, which satisfies both these requirements.

Inside the house

Coming inside to resume the planning of the interior part of the system, we find the next important feature is the panic buttons. The possibility of attack while at home by a bogus caller at the front door, or an intruder breaking in or walking in at the back while the door is unlocked, is unfortunately ever present. Panic buttons connected to the alarm system can sound the alarm even though it is not switched on, and so are ever ready for an emergency. Once triggered by a panic button, the alarm continues to sound until it is reset at the control box and the button is released by means of a key.

Any number of panic buttons can be installed and some thought should be given to their location. An obvious place for one is near the front door so that it can be actuated if a caller tries an attack. However, the attack may push the occupant backwards, and if the button is right by the door, it could then be out of reach. It may therefore be an advantage to place the button a little way back from the door.

It should not be installed where it could be accidentally set off, yet it should be easy to find in an emergency. It should be so placed that the hand would go to it without needing to look. Some experimenting should be tried with the button held in different places to determine the optimum. It should be out of reach of young children, who may think it great fun to set off the alarm, but not beyond the reach of older children who may have to use it. Adult shoulder height would seem to be about right.

Another location requiring a button is in the main bedroom, perhaps by the bedside. This could be used if an occupant happened to be upstairs when an intruder broke in on the ground floor. If near the bedside, it could serve if someone was resting during the day or ill in bed when an intrusion occurred.

The house layout will determine which if any other positions may be required. Some possibilities are: in the kitchen, by a home-worker's desk, or anywhere the occupant could be isolated from the principal buttons by an intruder forcing an entry.

One general point regarding panic buttons for domestic use could be made here. Most operate in conjunction with a 24-hour closed loop and so the contacts are normally closed. They tend to be over-engineered, large and ungainly, and to need a key to reset. These features are undoubtedly of value for maximum security in business premises where there is a possibility of previous tampering as a prelude to a planned attack. But this is not the case in the home.

Some feel that in the home such buttons are unattractive and conspicuous, and can spoil the look of any decor. It is understandable why some owners are reluctant to have them appearing here, there and everywhere. Yet if the control panel offered a latching normally open panic circuit, there would be no need to use a special obtrusive button that requires resetting, and an ordinary bell push would do the job. The push could be inconspicuously mounted anywhere, such as on the side of a door or window architrave, and there could be little objection to its appearance.

This then is another example of 'friendly security'. Unfortunately, the latching normally-open panic circuit is a rarity among control panels. However, it is a feature of the Sureguard system described in Chapter 14.

Internal sensors

These are our second line of defence to warn if the perimeter is somehow breached, or in some cases may augment it if it proves impractical to provide complete perimeter protection. We have already considered one example in the case of the bay window, where a pressure mat can be laid if it is not practical to wire all the opening windows.

Other places where pressure mats can be used are: in the hall; on a stair tread; in front of prize and trophy cases; in front of the main-bedroom dressing table; and so on. As pointed out in Chapter 5, a cat or a small dog would not actuate a pressure mat, although a large one may.

The PIR sensor can be used to protect large rooms having a number of possible entry points. An example is the front-to-back living room created where the centre wall has been removed. Here there are windows at the front and back, two or more doors, and possibly french windows. All these should be wired with sensors, but a single PIR detector will back these up.

It should be mentioned here that some PIR devices are sold as a complete self-contained system that consists merely of a single PIR detector and a sounder. These are claimed to give full home security. As they can only protect one room it is obvious that this is not so. One would be needed for each room, and even then the sounder would be unlikely to be heard far outside, if at all. An intruder could of course easily silence it by destroying it.

These devices may be useful for temporary portable security such as in a hotel room or a holiday flat, or as an independent back-up for an installed system, but they should not be solely relied upon for home protection. In fact their use may lull the owner into a false sense of security.

Control and zoning

Next on our planning agenda is what type of control unit to get and where to put it. An anti-tamper loop is not required, and neither is a timed exit circuit if any of the alternative suggestions are to be used, but most panels have them anyway. The basic requirements are: closed and open detection loops, panic circuit, and bell timer. With some panels pressure mats are connected across the detection loop and anti-tamper circuits, so the latter would be required even if not used as such. This arrangement would take the place of the open detection circuit. Sometimes the panic buttons are connected to the anti-tamper circuit instead of having their own separate circuit. An auxiliary switched supply will be required if PIR detectors are to be used.

The type of main control for the panel depends on personal preference. The principal choice is between key operation and a key-pad of numbers like a calculator. The key-pad would seem the best choice for those whose key-rings are already too heavy, while a conventional key would be safest for those with a poor memory for numbers!

Many home system control panels have just a single zone, which may be sufficient, but there is an advantage in having a two-zone or even larger system. Many burglaries are committed by the intruder entering at the back of the house and slipping upstairs to take jewellery from the bedroom while the family watches TV at the front. Having the back of the house on a separate zone, which could be switched on when the family is occupied at the front, thus makes good sense.

Another use for zoning could be the wiring of all upstairs sensors on a separate zone. These would be off at night but switched on when the house is empty during the day along with all other zones. Either of these options could be entertained with a two-zone panel, but if both were desired, a four-zone unit would be necessary. The fourth zone could then be used for a garage.

The control box should be located within the protected area so that it cannot be reached without triggering a sensor. Yet it must be possible to leave the premises without setting off the alarm, after switching it on. Normally this is done by making use of the timed exit facility but, as we have seen, a more convenient alternative is to have a lock-switch on the exit door that either shunts the exit door sensor or remotely switches the system. Alternatively a really stout door with a couple of deadlocks or deadlatches enables exit door contacts to be dispensed with altogether with little loss of security and a big increase in convenience; protection is afforded by a strategically placed pressure mat in the hall.

Control panels are often fitted in the hall in full view, but there is no reason why they should not be located elsewhere. In fact it is not the best place for them. The important thing is that the unit should be concealed, as a few hefty blows with a crowbar or jemmy could soon put paid to it, or it could be levered off the wall thereby breaking its wiring and silencing the alarm. It is not always easy to hide the box in a hallway; but it could be put in a cupboard behind some carefully placed 'junk', or on a wall behind a clothes rack, with clothing hanging in front.

Much depends on the layout. Too remote and secure a location may create inconvenience when leaving and entering the premises, and result in the owner not bothering to switch on when popping out for a short while. This can be an expensive mistake, as many householders have discovered! Furthermore, if the timed exit facility is to be used, there must be sufficient time to get from the panel and through the exit door before the time elapses, so it cannot be too remote. As long as it is not actually conspicuous and inviting attack, intruders are unlikely to go searching around for it with the alarm sounding. So it can be located in any convenient position as long as it is concealed and not too remote.

An important point, too, is the power supply. Often with DIY installations the box is plugged into an ordinary electrical outlet socket on the end of a long lead. This creates the possibilities of the plug being pulled out to plug something else in, or the socket being switched off. Although there may be standby batteries, these will not last for long if the power remains disconnected for any length of time.

A special fused wall connector should be installed for the purpose which prevents inadvertent disconnection. If you feel doubtful about doing this it should be done by a competent electrician.

Public halls

As many readers will have connections with the running of church halls, community centres, village halls and the like, and also the second Sureguard alarm in Chapter 14 is designed for such, we will devote some space to considering the particular needs of these.

Public address equipment is often the target, but generally halls contain few if any valuables, though they are vulnerable to youths who think there may be some cash, cigarettes or alcohol on the premises. Of course there are always the vandals who delight in breaking in and causing damage just for the sake of it. High security is thus not needed, but all windows and doors should be physically secured as described previously.

Windows tend to get broken, and it is not a bad idea to replace each breakage with unbreakable clear plastic. This spreads the cost and effort gradually instead of doing them all at once, and concentrates just on those windows that do get broken. It should be noted though that the plastic is not quite as clear as glass, and scratches easily. If there is a high incidence of window breaking, this may be the lesser of two evils. Alternatively there is unbreakable glass, but this is rather expensive.

It is certainly worthwhile installing an alarm system, but if there are a large number of windows it may be more practical to opt for a couple of PIRs in the main hall. These should be mounted high, one each end of the hall, in a corner, which should give complete coverage. Any doors and windows that are obscured from public view should be wired, as should windows to ante-rooms.

If valuable equipment such as a sound system is kept at the hall, this should be physically well protected in a stout casing. In some instances, a brick-built enclosure has been used to good effect. If it is in a cabinet, the latter should be heavily constructed, locked and secured to the floor.

The main problem is likely to be due to cleaners, hirers and others forgetting to switch the system on when leaving. The ideal set-up is to have the control box out of the way, and the switching accomplished by a lock-switch on the main entrance. This is the case with the Sureguard. It features a buzzer which is fitted near the entrance. If a sensor is activated by a window left open when the key is turned, the buzzer sounds continuously. After locating and closing the offending window, pressing a cancel button on the control box resets the system. If all is well, when the key is turned in the lock the buzzer sounds for a few seconds and stops to indicate the system is on. Unlocking the door switches it off. The system is thus fairly fool-proof (or cleaner-proof).

Strangely, in view of the obvious need for such a unit, they are not easy to come by commercially and some compromise may be necessary. This most likely will mean no warning buzzer, so the alarm will sound if the door is locked with a window switch or other sensor active. As there also is no means of telling whether the system is working or not, a regular test would have to be made by the caretaker or other responsible person. The best course is to build the Sureguard.

The bell should be fitted high on the front of the building if, as most are, it is a single-storey structure. Another should be fitted high inside the main hall.

Guest-houses

While these strictly are commercial premises, many large private homes in tourist areas are turned into guest-houses for the holiday season, so a few observations may help. Actually, they can be one of the worst headaches from the security angle. They can be protected just as a private house during the off-season period, but when the season is in full swing there are guests wandering in and out at all hours and keys are handed out to all. There are likely to be valuables left in guests' rooms, although a prudent guest-house owner will warn against this. Really, a guest-house can be a burglar's paradise.

Any conventional alarm system is likely to come unstuck, with guests leaving bedroom windows open and even perhaps opening lounge windows. The main danger is that of an intruder slipping in during the day while guests are out, or perhaps when they are at dinner and the owner and staff are occupied. The bedroom locks fitted in the majority of guest-houses are child's play to an experienced burglar, while the use of a ladder and window-cleaner's outfit could easily disguise an entry by a window.

One system which can be suggested, though it is unconventional, is a practical method of protecting guests' belongings in the bedrooms. All bedroom locks could be changed for lock-switches which could only be locked from the outside. Internal bolts could be provided for security of guests at night. The lock-switch could be connected in series with one or two pressure mats near the window and dressing table, the whole being wired into a central alarm system along with all the other bedrooms. Each floor should be on a different zone so that an alarm could be quickly localized.

When the door is locked, which can only happen when the guest is out, the pressure mats are on guard to trap anyone entering through the window or forcing the door without unlocking it. When the guest returns the door is unlocked and the mats are deactivated. As the door cannot be locked from the inside, visitors cannot trigger the alarm inadvertently. Indeed there is no need for them to even know that an alarm system is operating in their bedroom, although some proprietors may choose to inform guests as an example of the management's concern for their interests.

It may be considered desirable to have a buzzer installed in the office or the owner's private quarters rather than an alarm bell, which could disturb and cause consternation to other guests, who may mistake it for a fire alarm. A discreet investigation can then be carried out without upsetting anyone else.

The system would need to be on 24 hours a day and so could be a separate one from the main system having external sounders, which would be used off-season. Alternatively, a control panel could be chosen which had 24-hour circuits that sounded a buzzer, as well as the normal switched detection loops that operated external sounders. Then the single unit would control everything.

In addition to intruders from outside there is the possibility of guests straying into private areas such as behind the bar. No doubt this would be locked when not in use, but pressure mats at strategic positions could serve as additional protection. These would need a separate system as it would have to be switched off at appropriate times; a very simple latching battery-operated buzzer circuit should suffice for this.

While most guests pay by cheque, there could be large amounts of cash on the premises, especially at weekends. Steps should be taken to ensure that this is adequately protected.

9 *Installing the system*

Having established good physical security, and then planned the system as outlined in the previous chapter and obtained the needed items, the householder must now actually install the alarm. Before installation, a further inspection of the premises should be made to determine where and how cable runs are to be made. There could be more than one possible route, and the one chosen should be not only the easiest but also that affording maximum concealment and protection of the cable.

This installation inspection may reveal that some slight modifications may be required in the original plan – a relocation of the control panel, for example. All the cables have to be run back to it, and the proposed position may cause major problems for some of the runs. When any amendments have been sorted out and noted, the work can start.

The control panel

The control panel must be fitted securely to the main structure of the building, i.e. not to flimsy partitions or other secondary structures. If it is mounted on a wooden panel that has access to the back, bolts should pass right through with a metal plate at the rear. If there is no such access, no. 10 woodscrews should be used of a length a little short of the wood thickness. If it is fixed to plastered brick wall, the brick must be drilled and plugged to take screws of length at least 1A in (38 mm). If the plaster is unusually thick, longer screws should be used. All the fixing holes provided in the back of the case should be used.

As noted before, the mains supply should *not* be provided via a 13 A outlet socket. It should be taken from a specially installed unswitched fused connector wired directly to the ring main as a spur. Once the position of the control box has been finally determined, the supply point is best installed in advance.

The bell

The first step in fitting the bell or any other type of sounder is to drill a hole through the outside wall to take the cable. A long no. 10 masonry bit fitted to a two-speed electric drill running on the slower speed is the best way of doing it (Figure 46). A drill with a mechanical speed change

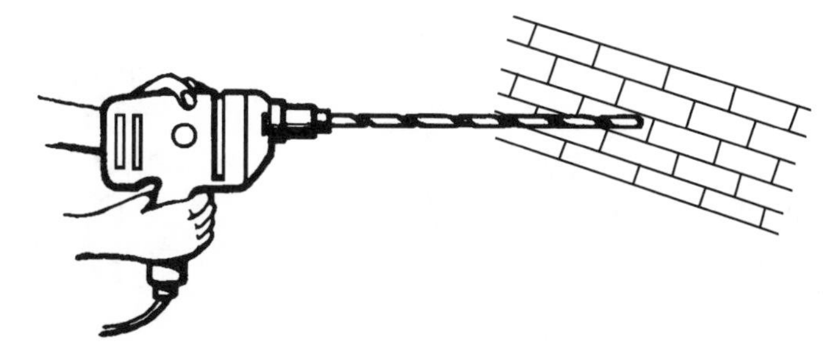

Figure 46 *A long no. 10 drill bit is the easiest means of drilling the wall hole for the bell wire.*

is better than the electronic type because it develops a high torque at the slow speed which the electronic speed control does not.

Some walls in older properties are too thick for even a long bit to penetrate. While very long bits are obtainable, these are expensive and unwieldy to use when the length is not required, and are also rather difficult to accommodate in the average tool kit. An alternative is to fit a bit extension. This is a shaft having a female thread at one end to accommodate the male thread on the end of the bit. As an alternative, the necessary equipment can be hired.

As a general rule it is better to start the hole from the inside, as usually precise location of the inside of the hole is more important than that of the outside. Also drilling from the outside could dislodge an area of plaster inside or otherwise damage the decorations. If the hole is being brought out under the floor, which is recommended, then it would probably be better to drill from the outside, as it may be difficult to drill from between the floorboards and joists. Accurate measurements will have to be made though to ensure that the hole comes out at the desired place.

Now the bell box baseplate, or the bell itself if of the open variety, can be fixed. Unless the unit is to be mounted on wood such as under the eaves, the wall must be drilled and plugged. The holes should be into the brick or stone, not into the mortar. Rustproofed no. 10 screws at least 1A in (38 mm) long should be used. An open bell should have a rubber or plastic ring fitted between it and the wall to prevent the ingress of moisture; with a rough-surfaced wall the gaps should be sealed with a suitable sealer. With an uneven stone surface, a flat area may have to be made by chipping away the raised portions and filling the gaps with cement to prevent the possibility of levering the bell off.

Wiring the bell

Ordinary 5 A mains flex can be used for wiring the bell. This is preferable to what is commonly known as 'bell wire', which may have an

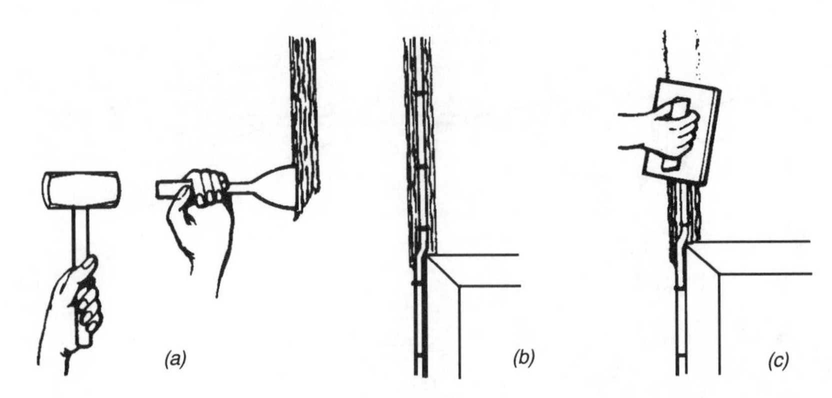

Figure 47 *(a) For cable down-drops from the floor above, cut a narrow channel in the plaster from ceiling to top of door frame. (b) Clip cable into channel and continue down side of door frame to skirting. Door frame saves having to make a long channel down the plaster, but for maximum security the bell wire should be buried in plaster all the way to the control box. (c) Finally, plaster over the wire.*

appreciable resistance over a long run and so result in a voltage drop. For high-security systems having a bell box with an anti-tamper switch, four-core cable will be required.

The security of the whole system depends on the bell wiring, so special efforts must be made to protect it over its whole length. It should be run well away from anything which could inflict physical damage.

Where high security is required, the bell wiring should be run in conduit or trunking, but not together with mains wiring. In domestic systems or where lower security is sufficient, the wiring need not be enclosed, but in all cases it should be concealed and if possible buried where it encounters plaster on the down-drop to the floor below.

This is not a difficult task. A narrow channel can be cut with a cold chisel and hammer; the wire is laid in it and secured with staples, and then it is plastered over (Figure 47). To cause the minimum spoiling of existing decoration, the drop can be run to coincide with a door frame; then it is only the plaster above the frame that is affected. The cable can be stapled down the side of the architrave for low-security systems or buried alongside it where greater security is needed. In most cases only a very narrow channel needs to be chipped out at the side of the wood, and the wiring can be lodged in it so that there is hardly any noticeable disturbance of the decor.

To pass the wire through a ceiling, a small hole can be made from underneath right into the corner between wall and ceiling with a long slim screwdriver. A little plaster can be chipped away from the wall so that the hole actually starts below the surface of the wall plaster (Figure 48). Thus when the wire is laid in the channel cut in the plaster it does not become visible where it enters the ceiling.

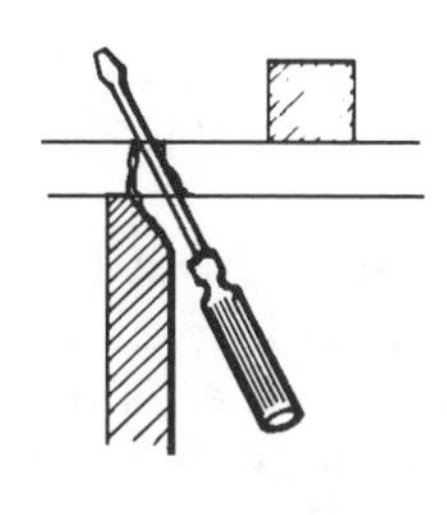

Figure 48 *A hole can be made in the ceiling with an old screwdriver (left); chip a little plaster from the top of the wall so that the hole is below the level of the plaster. If the screwdriver is pushed in up to the handle (right), the blade will mark the other side of the hole which may otherwise be difficult to locate.*

When the ceiling hole is made with the screwdriver, push the tool into the hole up to the handle and leave it there. Then the blade can be seen under the floorboards upstairs, and so the hole is easily located. If the blade is withdrawn it can be surprisingly difficult to find the hole from above.

The second exterior bell, if fitted, can be installed in the same way, and also the interior one. The bottom part of the bell wire run where it enters the control box is particularly vulnerable. All wiring entering the box should be buried in plaster and preferably laid in trunking. Fairly wide channels should therefore be cut in the plaster from the control box in the direction of the main cable runs, at the start of the job, so that all the system's wiring can be laid in as it is brought back. When all wiring is in place, the trunking can be capped, and plastering made good.

Sensor loops

Wiring the loops and installing the sensors are the biggest part of the job. A separate loop is of course needed for each zone. It is recommended that no more than ten sensors be connected to each zone to facilitate rapid identification of an alarm, whether false or actual. Four-core cable is required for high-security systems, one pair for the detection loop and the other for the anti-tamper loop. This is not necessary for domestic installations, but if two zones are to be formed and the wiring heads in the same direction for both, it may be practical to use four-core wire, a pair for each zone.

On the other hand, this will require joints where the circuits separate, and the four-core cable will have to be purchased in addition to a reel of two-core. So it may be better to use two-core throughout and bring them back to the control box, stapling them together. It all depends on the layout.

For sensors such as PIRs that require power, another pair is needed, making four cores. Suitable cable is 7/0.2 mm 1 A having a thickness of R in (3 mm).

Most of the sensors will be magnetic reed switches with matching magnets. The switch is fixed to the opening side of the door or window frame, while the magnet is fitted to the door or window itself. Some switches have lead-out wires while others have terminals. Joints to the former should be either soldered and insulated, or made with a junction box, so those with terminals are the easiest to install. Wherever possible use the flush-mounting variety for higher security. For metal or UPVC windows, surface-mounted sensors with leads are required and either soldered or junction-box joints made.

It is easy to make a mistake when connecting the cables to sensors, and a single error could take a lot of time to trace and correct later. Most cylindrical flush-mounting terminal-type magnetic switches in common use have five terminals at the back in a pentagonal configuration: the two with a silver finish are the switch terminals, while the three with a brass finish are blank for connecting the other wires and the anti-tamper loop.

Thus one of the incoming pair from the control unit is connected to one silver terminal and the second to one of the brass ones. One of the outgoing pair to the next sensor is connected to the other silver terminal, and the second outgoing wire to another brass terminal that is internally linked to the first brass terminal. The next sensor is wired in exactly the same way, the incoming pair being the other end of the outgoing pair from the previous one. With the last sensor, the incoming pair from the previous one is connected to the two silver terminals (Figure 49). If you are not sure which brass terminals are linked and have no continuity tester to check, connect both second wires together on the same terminal. Thus all magnetic switch sensors are connected in series to form a continuous loop. It must be noted that not all units are of this pattern, but the switch contacts should be clearly indicated.

Pressure mats have four wires several inches long coming from them. Two are for the anti-tamper circuit which is simply a single wire that goes into the mat and straight out again. These can be cut off close to the mat as they serve no other purpose. The four leads are identified by the insulation which is stripped off the ends of the switch wires but left intact on the anti-tamper ones.

With some control units the mats must be connected across the closed-loop and anti-tamper circuits, but with others a separate normally-open circuit is provided. A junction box is required for each mat, these being specially made for the purpose. All mats are connected in parallel, so the

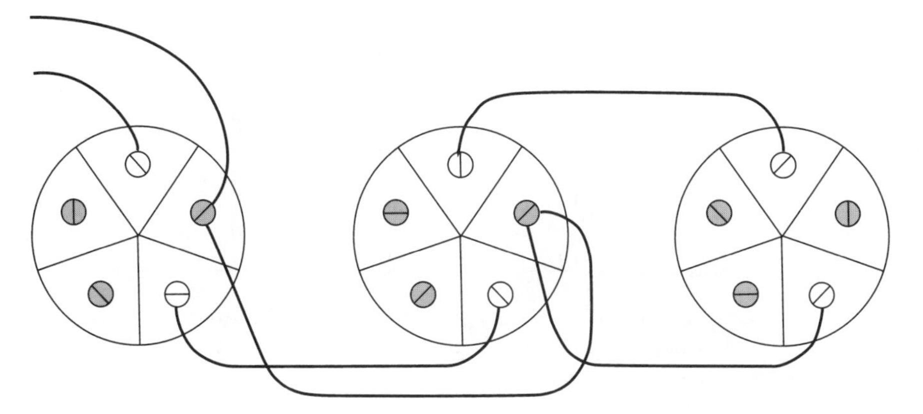

Figure 49 *Connection of five-terminal magnetic sensors. Silver finished screws are the switch terminals, brass ones are blank.*

Figure 50 *Connecting a wire to a screw terminal. (a) If the wire is given a clockwise twist, tightening will draw the wire into the centre. (b) If an anticlockwise curl is made, tightening will push the wire out towards the edge where it may come away.*

pair of wires from the control box is connected to the same terminals as the pair from the first mat; another pair of wires from a linked pair of terminals (or the same pair) goes to the next junction box; and so on.

Here is a tip when connecting wires to screw terminals. Wrap the wire in a clockwise direction around the screw; then when the screw is tightened the wire is pulled in towards the centre. If it is wrapped anti-clockwise it will tend to be pushed out (Figure 50).

Junction boxes should be made as inconspicuous as possible, being mounted low on the skirting board, behind furniture if convenient. There is no need to be too fussy about this though, as they are unlikely to be tampered with in a domestic situation, and an intruder would not be interested in it after he had stepped on the mat and set off the alarm, even if he knew what it was.

Running the loops

There are two possible ways of wiring a sensor loop: one is to run the wire at floor level along skirting boards; the other is to take it above the ceiling beneath the floorboards of the storey above. The latter may be

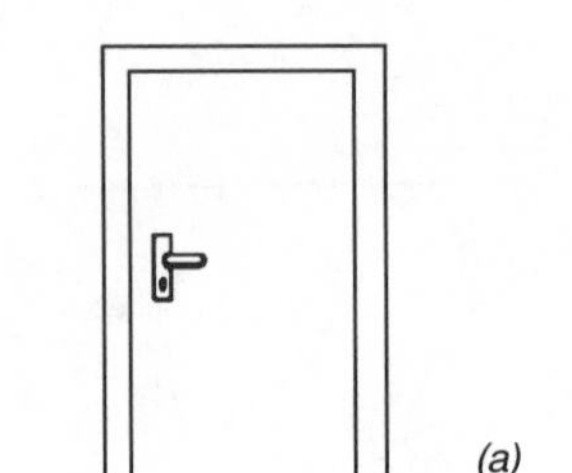

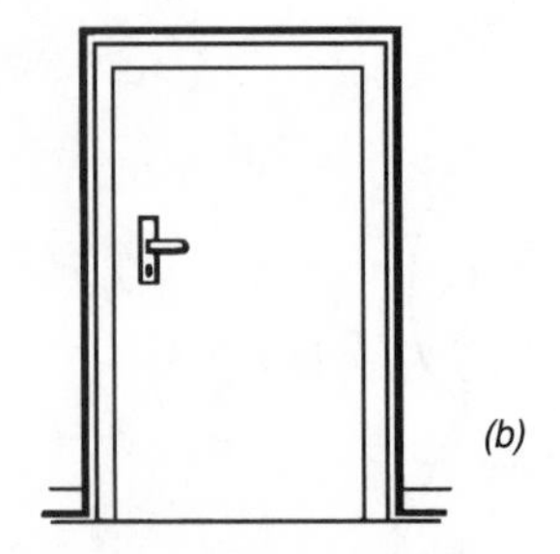

Figure 51 *Crossing a door can be a problem. (a) If there is a carpet extending through the door or a fixed draught excluder, wiring can be taken underneath. (b) If the floor is bare, wiring must be taken over the door frame.*

easier if there is not much furniture and the floorboards are easy to raise. If they had to come up to run the bell wire it may prove most practical to lay in the loop cables at the same time. Loops serving upper and lower floors could also be put in at the same time between ceiling and floorboards. These are the points which can be determined during the previously mentioned installation survey.

Running the wiring along skirting boards causes least disturbance and is probably the most practised method, but it offers less security than under-floor distribution to which there is no access for would-be tamperers. This may be of concern with high-security installations but is less so with normal home systems, although the possibility of accidental damage must be considered.

The main snag with the skirting method is crossing doorways. The wiring has to pass either under the door space, or up and over the architrave (door frame), which means greater exposure of the cable and so reduced security. If there is a threshold draught excluder or carpet joining strip, the cable can be concealed under it. Otherwise it could pass under carpeting providing it was secured at both sides and covered with shallow bevelled channelling (Figure 51).

The location of the sensors on the door or window is determined by which of these two wiring methods is used. If the wiring is taken around the skirting, the contacts will be low near the bottom of the door, but if the wiring comes down from the ceiling, the sensors will be fitted to the upper part (Figure 52).

Fitting door switches

A E in (19 mm) hole must be drilled in the door frame of sufficient depth to accommodate the body of the switch and also some surplus cable. This enables the unit to be withdrawn if it should need to be checked or

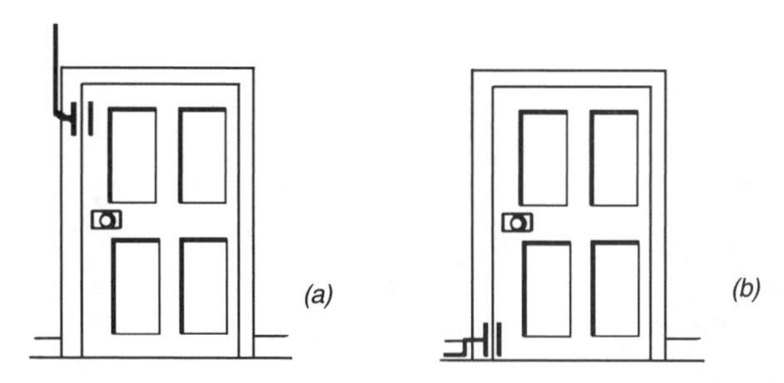

Figure 52 *Position of sensors depends on the run of wiring. (a) If dropped from the floor above, sensor can be near the top of the door. (b) If run along the skirting, sensor can be low.*

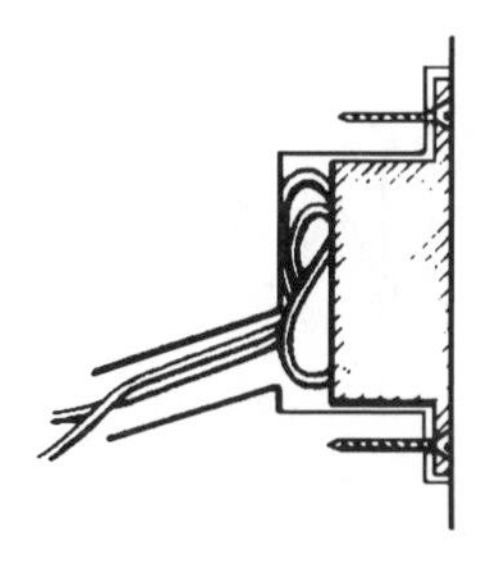

Figure 53 *Flush sensor in position in cavity. Surplus wire needed to enable withdrawal for future examination is stored at bottom of cavity which should be deep enough to accommodate it.*

replaced in the future (Figure 53). A neater job results if a shallow cut-out is chiselled to accommodate the flange and so make it truly flush. This is not strictly necessary though as the flange is very thin.

A hole must then be drilled for the cables from the side of the architrave to the bottom of the switch hole. Architraves in many older buildings are so wide that a normal wood drill will not penetrate, especially if the hole must be run on a slant. Other than obtaining extra long drills, two holes must be drilled, one from each end to meet in the middle.

This sounds very difficult, and indeed if they are drilled in a straight line the chances of them meeting are slim. The chances are considerably improved if both are drilled at a slight downward angle so that they form a shallow V. This gives a certain tolerance to the vertical angle of the second hole; if it is drilled at a slightly different angle it will still intersect with the first (Figure 54). The first hole from the cavity can be large to further improve the chance of intersection, as it will be concealed, but the second should be only wide enough to take the cables.

When the switch has been fitted, the mating magnet can be mounted exactly opposite in the door or window. The magnetic field will extend over normal gaps, but excessive space may require special higher-powered magnets, otherwise there may be false alarms due to the field being insufficient to hold the switch.

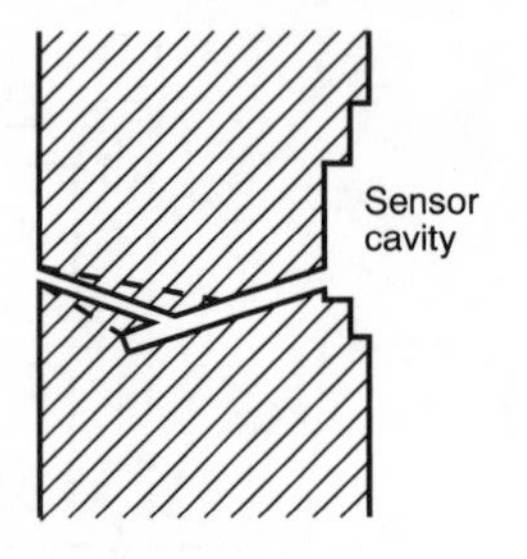

Figure 54 *Making two holes meet in a wide door frame. The first hole from the sensor cavity is made with a large bit at a slightly downward angle. The second hole from the opposite side is of smaller bore, also slightly downward. Dashed lines show the angle tolerance; the hole can be drilled anywhere between these angles and still intersect with the first one.*

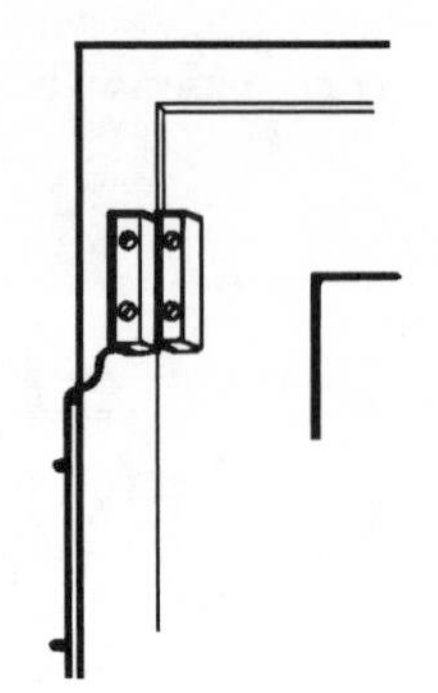

Figure 55 *Surface-mounted sensors screwed to door and frame.*

Surface contacts should not be used for high-security systems other than on metal or UPVC window frames, as they can be defeated by previously attaching a small bar magnet which keeps the contacts closed. This ploy is not detected by the anti-tamper circuit. Surface contacts can be used for most domestic systems, where such a possibility is remote, but they should always be fitted to the inside of the door (Figure 55).

Window foil

For highest security each area of glass should have two separate foil runs, one connected to the 24-hour alarm circuit such as used for panic buttons, and the other to the anti-tamper circuit. If either is broken at any time, whether the alarm is on or not, an alarm will be triggered. It will also sound if the foils are short-circuited together in an attempt to defeat them. If the control panel does not have a separate 24-hour circuit, a two-pole arrangement can be achieved by using the normal loop plus the anti-tamper circuit. In this case, a short-circuit will not trigger the alarm immediately, but it will when the system is switched on, so warning of previous tampering.

Also for highest security, adjacent runs should be no more than 8 in (200 mm) apart. Runs that keep close to the edge of the glass are not very secure, and some coverage of central areas is essential.

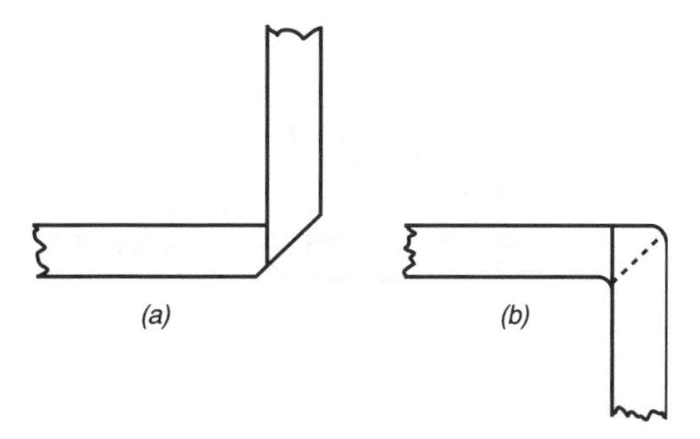

Figure 56 *Forming a corner with window foil: (a) first fold the foil in the opposite direction; (b) then fold back to form right-angled corner.*

For ordinary domestic systems, this degree of security is unnecessary owing to the unlikelihood of previous tampering. Foil will be used only on those permanently closed windows which are in a vulnerable position. For small windows a horizontal strip a few inches from the top and from the bottom will in most cases suffice; for larger ones, a rectangle of foil a few inches from each side should be formed. The very presence of foil, which can be seen from the outside, is a deterrent to an intruder. There is little point in him trying to remove or cut out a window if the alarm will be immediately triggered.

If the control unit has an anti-tamper loop facility, this should be used for window foil as it then will be protected 24 hours a day. Failing this it should be connected in series with the closed loop and so be on guard when the system is switched on.

The first step is to clean the window with methylated spirit to remove all traces of grease. The window should also be dry and free from condensation when the foil is applied, so if the job is done in the winter, some form of heating should be placed in the window bay for at least half-an-hour beforehand.

Getting the foil to lie straight is not as easy as it may seem, and a crooked run can look very untidy. The solution is to lay masking tape on either side of the path of the run, which will serve a further useful purpose later. It is then comparatively easy to lay the foil in the space between the edges of the masking tape.

Corners are the next problem. Do not attempt to join lengths of foil: the adhesive between them will prevent electrical contact. To turn a corner, first fold the tape in the opposite direction, making a 45° crease. Then fold it back in the required direction and press down (Figure 56).

Do not run foil over any cracks or butt joints in unframed glass, as it will surely break there. Foil is fragile – it has to be to fulfil its purpose – so it needs care in handling. Connections are made by means of adhesive

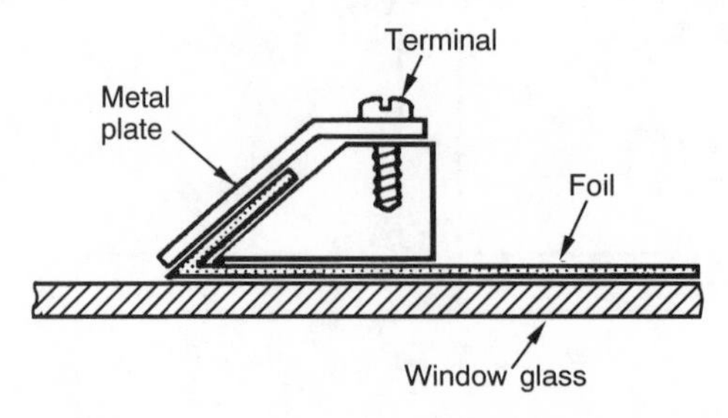

Figure 57 *Making off foil with self-adhesive terminal block. Block is fixed to the glass over the foil which is bent back on the block and clamped by the terminal plate.*

terminal blocks which are fixed to the window glass over the tape, the end of which is brought up and clamped between the block and a metal plate. A screw in the plate connects the wire (Figure 57). Do not leave any loops or raised portions of foil, as these could be broken by window cleaning.

It is preferable if possible to mount the connectors near the top of the window as they then are unlikely to become waterlogged by condensation running down the glass. It would also be difficult for them to be tampered with unobserved if fitted there, by children if not intended intruders. The worst place is at the bottom. Connectors should be mounted within 4 in (100 mm) of the window frame.

Finally, the foil should be painted with a coat of clear varnish to protect it against corrosion caused by condensation. This is where those strips of masking tape prove their worth, as the varnish extends only a millimetre or so beyond the edge of the tape and ends in a clean straight line. The job is completed by stripping off the masking tape when the varnish is dry.

Vulnerable ceilings can be protected with hard-drawn lacing wire which should be zigzagged across the area and secured by staples at the sides. It can be prevented from sagging by applying sticking tape at intermediate points across the runs. This can then be papered over with an embossed paper. No unsightly protuberances will result as they would if staples were used for support. The ends are connected with terminal blocks. Like the window foil, the wire can be taken to an anti-tamper circuit for round-the-clock protection, or to the normal closed loop. This wire is supplied on 100 m drums, so quite large areas can be economically protected.

Vibration, impact, inertia sensors

Not normally part of a domestic system, these devices have their uses for special circumstances such as high-security installations and the protection of outdoor perimeters. They should be installed according to the maker's instructions which may differ from model to model.

Vibration sensors are the simplest and are just connected in series with a 24-hour detection or anti-tamper loop. They must be installed vertically the correct way up and, after installing, the sensitivity must be adjusted by means of a set-screw. Where there is a lot of normal vibration such as on main roads or near machinery, the sensitivity must be set low to avoid false alarms. In a more peaceful environment the sensitivity should be set high.

Impact detectors can be mounted in any position. They are powered and so need a six-core cable. Each unit has its own self-contained analyser to provide rejection of incidental impact shocks and to set the sensitivity.

Inertia detectors can be either flush or surface mounted. They can be used in window frames at any convenient spot instead of magnetic switches. For fencing they should be fixed at 10 ft (3 m) intervals, while for walls the spacing can be 12–16 ft (3.5–5 m). For the latter they should be mounted away from floors and ceilings and abutting walls. All sensors must be mounted vertically, and an arrow is usually inscribed on the unit to indicate the correct way up. They require an analyser which counts the impulses and their severity, registering an alarm only if these exceed a certain pre-set value.

A number of detectors, up to 15 with some models, can be connected to each analyser, but where different sensitivities are required for different areas, multiple-zone analysers are needed. Each sensor is wired with a four-core cable, of which one pair is a series detector loop which is connected to the analyser, and the other is the anti-tamper loop which goes back to the control panel. With some models the final sensor has an end-of-line series resistor so that any short-circuit across the loop increases loop current, which is detected by the analyser and triggers the alarm. The analyser is connected to a 24-hour detection loop at the control panel, and also to a 12 V power supply. It is best to locate it near the control panel. With fencing, the sensors need to be fitted in electrical conduit boxes if not already suitably enclosed, at the upper part of the fence, and the wiring is run along the fence in conduit.

Active infra-red detectors

With active infra-red systems, a 12 V supply is needed for both receiver and transmitter, but these need not be from the same source. The receiver can be supplied from the control unit in the same cable as the detection loop, but it may be more convenient to obtain the power for the transmitter from a small power unit located near it. This could be so if the beam is required to protect a space between two buildings. In such cases it is necessary to ensure that the transmitter is always powered.

An anti-condensation heater is usually included in both units, which requires a 12 V supply. Transmitters need 25–100 mA, receivers around 30 mA, and heaters up to 250 mA. The control panel may not have sufficient auxiliary power to provide heater power and a separate supply may be necessary. Separate cable should be used, as will be shown later.

If used to protect a perimeter fence, the units should be mounted several feet inside the fence so that the beam cannot be cleared by jumping over the top. At least two vertically aligned beams should be used in parallel so that birds flying through one will not cause a false alarm. Alternatively, a double-knock analyser could be fitted which ignores a single short break. While external infra-red units will generally operate in sunlight, it should not be allowed to fall directly on the active surface of the receiver.

The use of mirrors to bend the beam and so cover two or more adjacent sides of an area is frowned upon by some authorities. The beam is reduced some 25 per cent by a reflection from a clean mirror, but much more from one obscured by condensation. As a mirror has no demisting heater, the possibility of dew causing false alarms is a real one.

Passive infra-red detectors

For indoor use, the passive infra-red is superior to the active beam-breaking system. PIRs are now the most commonly used space protection devices. Many installation firms are using them in place of magnetic switches because they mean less work, but this practice is questionable. They allow the intruder to get inside before the alarm is triggered. The perimeter should be the first line of defence, and space protectors should be the back-up or should fill in where it is impractical to fully protect the perimeter. Apart from the light-control PIRs which are independent self-contained units, those used with alarm systems are either latching or non-latching types.

The relay contacts in the non-latching PIR sensor are open-circuited when the device is activated by a moving heat-radiating body and the detection loop is broken, just as with magnetic door contacts. When movement ceases the relay contacts return to their former position. In addition an indicator light comes on, and goes off when the movement stops. This is known as a *walk-test light* because it enables the sensor to be tested for range and coverage by walking within the protected area with the alarm system switched off. (The needed 12 V supply is available from the control panel when it is switched off.)

With latching PIRs, the light stays on after the alarm source has been removed and the panel switched off, thereby identifying which sensor was responsible. This is particularly important when checking out a false alarm, so if more than one PIR is to be used on the same zone they should be of the latching type for this reason. An extra wire is needed to supply a switched 12 V from the control panel when it is switched on. Non-latching PIRs are wired with six-core cable for the detection anti-tamper circuit and the power supply, while latching models need eight cores, one of which is unused. Where the anti-tamper circuit is not used, as in domestic systems, the cable requirements are four and six core respectively.

Siting PIRs

Care must be taken in siting the PIRs to achieve the desired coverage and also eliminate the risk of false alarms. In Figure 33 a number of typical coverage patterns were illustrated and a sensor should be chosen that gives the nearest pattern to fit the particular situation. In most case the horizontal coverage is that of a right angle. This suggests that the best position in any rectangular area is in a corner. If mounted midway along a wall there will be blank areas on either side. If such a position is unavoidable, there are models that have a wider side coverage. Alternatively, the PIR can be sited on a wall facing the access point. Then the blank areas will not matter as they cannot be reached without passing through the active regions.

Among the coverage patterns shown in Figure 33, three are worthy of note. One has a totally horizontal distribution and is intended where pets or other small animals could wander into the area (Figure 33a). It should be mounted waist-high. In theory it is less secure than other types because a human could crawl underneath the detection region, and it is at a level at which it could be easily be tampered with. However, the intruder would have to be aware of its presence and detection pattern, which is unlikely.

Another unconventional pattern is the long narrow one (Figure 33c). The range is up to 120 ft (36 m) but it is confined to a narrow pencil-like pattern with three vertical regions. This is ideal for long corridors and passages.

The third pattern is like a conventional one, but turned on its side (Figure 33b). It can be used to protect tall narrow areas such as tall windows and stairwells. The range is shorter at 34 ft (10 m) but is ample for most uses.

Range is another factor to be considered. The usual ranges are from 30 to 50 ft (9 to 15 m). For most domestic indoor applications this is ample. As nearly all PIRs aim downwards, the height of the unit governs the

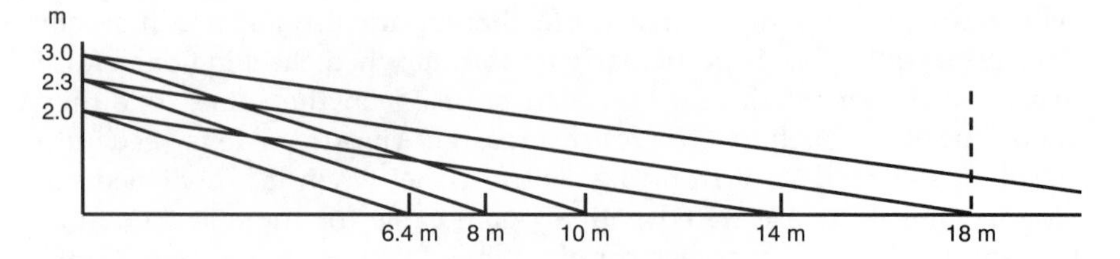

Figure 58 *Plot showing how the range of a PIR depends on its height, at normal mounting angle and also tilted forward 10°.*

range (see Figure 58). A lower position shortens the range, while a high one achieves the maximum: 6–9 ft (2–3 m) is the usual. Most sensors have two or three vertical detection regions to cover distant and near areas, as shown in the illustration. While a high position increases the range it also widens the blind area immediately below the detector. The best height will thus depend on whether the long range or the short is the most important for a given situation.

Some units have an adjustable tilt facility, either of the case or of the reflector inside. This enables the desired coverage to be obtained at any height. A high position is preferable to make tampering more difficult, so this facility is a useful one.

Where there is a possibility of objects near the limit of the range causing false alarms, a sensitivity control in some models permits the range to be shortened.

The detector should not be pointed towards a forced air heating duct as the movement of warm air could trigger it. It is best to site it so that it aims in the opposite direction. Other forms of heating usually cause no problems. It also should not be directed towards a south-facing window. Glass attenuates infra-red radiation, so penetration by body heat from persons outside would not affect the sensor, but movement of very warm outside objects heated by the sun could. Apart from these there is little that can create false alarms. Direct sunlight should be prevented from entering the lens as this could damage the sensor.

It should be noted that PIRs are not tamper-proof. In most domestic situations this will not be a hazard, but if persons such as builders, decorators, cleaners, social workers etc. who could be in collusion with potential intruders have access to the protected area, they could easily incapacitate a PIR detector by sticking paper or tape over the lens. It would be likely to pass unnoticed and would not be revealed as a fault when switching the system on. So an anti-tamper circuit for PIRs is rather superfluous as no intended intruder who knew his onions would fiddle with the cover or the wiring when there is such an easy way to defeat the device. This is why they should be mounted above normal reach.

Incidentally, it is also the reason why total reliance on PIRs is unwise; they should only be used to back up perimeter sensors.

Ultrasonic detectors

Being more prone to false alarms than PIRs, these are far less popular than they once were. The principal difference in the detection characteristic is that the PIR is most sensitive to sideways movement across the detection field, whereas the ultrasonic device is more sensitive to movement to and fro in it. However, while this is so, both are quite sensitive in the other direction and there is little reason to choose one over the other on this score alone.

Wiring is similar to that for the PIR detector: a pair for the 12 V power supply, a pair for the detection loop, and a pair for the anti-tamper circuit if used. Like the PIRs, some ultrasonic sensors have a latching facility which needs an extra wire.

Siting and setting up are very important to minimize the risk of false alarms. Environments where there may be high-pitched sounds such as from steam escaping, gas heating, some types of machinery, and brake squeal from passing traffic, must be avoided. The detection pattern is not so clearly defined as that of the PIR, so the device must be aimed well away from anything that is likely to move, including curtains.

The better ultrasonic sensors include some discrimination against small and random movements such as minor air turbulence and insect flight, but they cannot entirely distinguish between different types of movement. A walk-test indicator is provided which flickers on when it detects movement and stays on for a few seconds when the movement is sufficient to cause an alarm.

A sensitivity control is provided, which should be set to the minimum required to produce an alarm when there is movement at the boundary of the area to be protected. Any higher setting will increase the possibility of false alarms. However, there is little that an ultrasonic detector can do that a PIR will not do better.

Exit circuit

Having previously decided which type of exit arrangement to use, we now have to look to practical installation details. If at all possible, only one sensor should be actuated by anyone leaving the premises after switching the system on. With a timed exit the time should not be set to more than two minutes.

Care must be taken, when siting space protection devices such as PIRs

Figure 59 *Key-switch for mounting in exit door.*

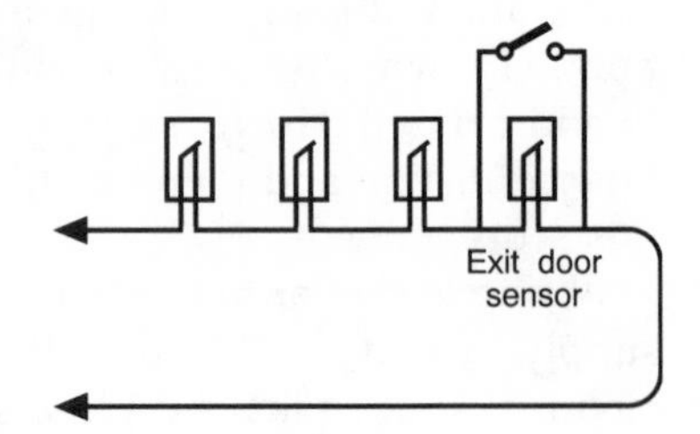

Figure 60 *Bypass switch connected across the sensor of the exit door. This can be either a key-switch or a lock-switch.*

in areas near to the exit route, that they cannot be actuated by someone on the exit route. It is best that space protectors are not used as the exit sensor.

In most cases the exit sensor will be the magnetic switch fitted to the exit door. The choice here is between wiring it separately to the timed exit circuit; connecting it into the normal detection loop and fitting a shunt switch; and fitting a remote control switch. When wired to the timed exit circuit, the cable should be run straight back to the control box with no other connection made to it.

A shunt switch can be either a key-operated switch fitted to the exit door or frame if wide enough (Figure 59), or a switch incorporated in a security deadlock. The latter has the advantage of requiring only one key. Lock-switches usually are of the single-pole double-throw (SPDT) type, commonly called changeover, and offer either opened or closed contacts when locked.

When used as a shunt switch, the lock-switch is connected so that when the door is unlocked the switch short-circuits the sensor (Figure 60). Wiring from the lock should be taken up the back of the door, and for maximum security covered with trunking that is well secured to the door. It is taken across to the frame by means of a flexible loop (Figure 61). These loops are made for the purpose and usually have four cores of which two can be ignored. The loop should be terminated in junction boxes, some being supplied with boxes already fitted to the ends. The wiring from the magnetic switch is also taken to the box on the frame, as is the wiring to the rest of the loop.

With a remote-controlled unit, the whole system can be switched on

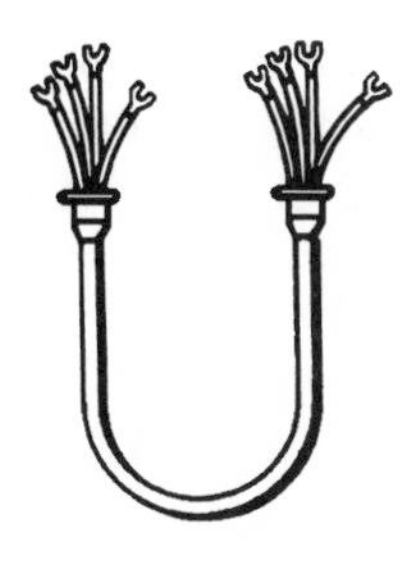

Figure 61 *Four-core loop for connecting door switch to wiring on door frame. Wires consist of flat wire wound on cord for high flexibility.*

and off from the lock-switch on the exit door. No operation is therefore required at the control panel itself other than for testing. This is especially useful for public halls which are hired out for various events and where the switching on of a security system could often be neglected. In this case, if the door is locked the alarm system is switched on, so the closed pair of contacts should be used. If a warning buzzer is included with the system, this must be mounted on the door frame so that it can be heard from outside, and wired via an extra pair of wires in the cable to the control panel. The Sureguard remote-controlled unit described in Chapter 14 uses all three contacts of the door-lock changeover switch, so three-core wire must be used, or four-core ignoring one wire.

Panic buttons

These are normally-closed switches connected to the 24-hour detection loop or, if the control panel does not have one, the anti-tamper loop. Although the loop circuit latches electronically in the control box, panic buttons have also mechanical latching which can only be released by means of a key. Thus the panic buttons for a loop are all connected in series, the same as the magnetic switches, but to the 24-hour or anti-tamper loop.

The buttons will be sited according to the predetermined plan, which will be wherever a personal attack is possible. Near front doors, perhaps in the living room, and in bedrooms are the obvious points. They should be easily reached without fumbling, yet not be vulnerable to accidental operation. Some trial positioning with persons likely to use them in an emergency is worth while to get the location dead right.

In spite of their large and formidable appearance, many panic buttons use a reed switch which has a low current rating. They should therefore not be used to switch current to a sounder direct. Some though are heavy duty rated and can pass sounder currents; in addition they are SPDT and so can be connected in either a normally open or a normally closed mode. These can be used in a simple panic system independent of the main alarm, and are wired as normally open just like an ordinary bell circuit. The mechanical latching ensures that, once pressed, the alarm keeps sounding. Wiring to the button should be protected or concealed as the alarm would stop if it was damaged.

Connections

The importance of making sound and lasting connections, especially in the detection loops, cannot be emphasized too much. As there usually are quite a number of connections around a loop, and there is more than one loop in each system, the chance of making a bad one is quite likely, especially for a non-professional installer. Yet just one poor connection in a whole system can produce false alarms that are very difficult to trace, cause a lot of trouble, and so undermine the effectiveness and credibility of the whole installation. We will here examine some of the chief causes of poor connections:

1 A common occurrence is that many of the strands in the wire bundle are partly cut when the wire is stripped of insulation. These can part when the wire is screwed down, leaving only one or two strands actually connected. They too can break as the sensor is manipulated into position, and although they may make touching contact, vibration, temperature changes or other causes can make them part momentarily, so producing intermittent faults that are very difficult to trace.
2 Stray wires can escape from the bared bundle and short-circuit to the adjacent terminal. This is less likely with sensors having raised insulation around the terminal, but it does happen.
3 An anticlockwise wire wrap around a screw terminal can force the wire out when tightened (see Figure 50). Even when laid clockwise the wires can spread out from under the screw head during tightening yet appear to be well secured. Sometimes only a single strand or so remains under the head, and this can be broken when the sensor is pushed into place. Contact may be made, but only a touch contact.
4 Another possibility is that part of the insulated wire is included under the head; so, being thicker than the stripped portion, it holds off the head from pressing hard on the bare wire. The result is a loose contact.
5 Terminals in which the screw end grips the wire may not suffer from some of the problems of screw head terminals, but they have some of their own. A sharp burr on the end of the screw can sever most of the wire when it is tightened.

6 Also, the wire may not penetrate far enough into the barrel of the terminal to be gripped by the screw. A common reason is that the screw was not first unscrewed sufficiently, and so obstructed the entry of the wire. When screwed up tight it may just pinch the end of a strand and so seem to grip the wire. Later, it comes loose.

It can be seen from this list that the chance of making a bad connection is by no means slight, especially when working in poor light and in cramped conditions.

The following tips, if rigorously observed, should eliminate the problems:

1. Use a portable free-standing lamp. This will give light where it is wanted and leave both hands free.
2. Use special wire-strippers, not pliers or side-cutters. There are a number of inexpensive yet excellent tools now available that take off the required amount of insulation cleanly, and without damaging the wire. They are also easier and quicker to use. They are well worth it to ensure freedom from false alarms and other problems in the future.
3. After stripping, twist the bared strands in the same direction as their natural twist before connecting, thereby preventing stray wires from escaping.
4. Use a clockwise wrap under the heads of screw terminals, pulling the wire close to the screw thread, but trim off the excess and avoid laying it over the first part. This creates a high spot that prevents the screw head gripping the rest of the wire, so reducing the contact area.
5. Check that there is a millimetre or so of bare wire exposed outside the terminal to ensure that no insulation has been trapped under the head.
6. With the barrel type of terminal, look into the end when loosening the screw to see that it is almost fully out and offers no obstruction to the wire, but not so much that it falls out and gets lost! When screwing down, hold the wire gently with the other hand to keep it in place. It is usually possible to feel the wire 'settling' under the pressure of the tightening screw, thus indicating that it is firmly gripped.
7. When the connection is complete, give a slight pull to ensure that the wire is fast. You may be mortified at the number that come apart!
8. Finally when all connections on the sensor are made, examine each in turn with a pocket magnifying glass in a good light. It is surprising what potential faults this can reveal, and it is well worth the extra time spent.

Some readers may feel that the above list is rather elementary and so may be inclined to be dismissive. However, even professional installers sometimes make bad connections, and it only needs one in a closed loop to produce unaccountable false alarms that can prove a real headache. Take your time over each connection and satisfy yourself that it is not less than perfect.

10 Faults and false alarms

The most common fault encountered with alarm systems is the false alarm. The causes in order of probability are:

1. misuse by user
2. incorrect siting or setting of space detectors
3. excessive vibration, wind, storm or other unusual physical actuation of sensors
4. wiring faults
5. poor alignment of magnetic switches
6. control panel fault.

Not to be overlooked is the possibility of a thwarted attempt at entry which has left little trace. For example, a hefty shoulder charge could move the top of an outside door sufficiently to actuate its sensor, but if the lock held, it would spring back into place. The alarm may be triggered but no trace of the attempt will remain.

The design of the detection circuits is such as to minimize this sort of thing happening because they ignore momentary breaks in the detection circuit of between 0.2 and 0.8 seconds. However, if a panel was designed for the faster response, such an event could trigger it.

When investigating an apparent false alarm, the first thing is to check which zone was triggered if the control unit has more than one zone. Most multi-zone panels have a latching indicator to show this and it eliminates at least half if not more of the system from suspicion. If the affected zone has sensors on outside doors or windows, carefully examine for any attempts at entry. Look for such things as fresh indentations in woodwork caused by a jemmy, or bruised paintwork, scratches, scuff marks or dirt on window sills caused by shoes, or footprints in soft earth.

If an outside door on the circuit has inside bolts as well as a lock, were they bolted? A door bolted top and bottom cannot be sprung inwards away from the frame sufficiently to affect a magnetic sensor, but an unbolted one could. Thus eliminate possible outside attack before assuming a false alarm.

Misuse, human error

Most false alarms are due to these, the exit being a notable example. With a timed exit, someone may have taken too long to leave after the panel is activated. Or, perhaps they remembered something that had to be collected in an adjacent area and so deviated from the exit path, thereby triggering another sensor.

These may not be serious as they are obvious and the persons concerned can in most cases deal with the alarm there and then, switching off and rearming the system. If the installation is connected to a telephone dialler, though, a full alarm would have been notified to the number dialled.

Another human error alarm could be caused by an object such as a chair placed on a pressure mat. If light, it may not have an effect immediately, but after an hour or so may compress the foam sufficiently to make contact. If any mats are connected to the affected zone, check that all are well free of obstructions, and question other family members whether anything has been moved since the alarm. The offending object might have been moved aside.

Perhaps the most common of this type of false alarm is caused by failure to secure windows or doors. They may be shut, but if not properly secured may be moved by wind or vibration. An example of one actual case demonstrates the problems this can cause.

A church hall was wired with magnetic switches on some interior doors as well as the usual perimeter ones. One door, which was sprung to keep closed, led to an ante-room which contained an access hatch to gas-fired heating equipment. The equipment was ventilated by a grille to the outside. Over a period, false alarms occurred that could not be traced, in addition to a real break-in.

Eventually the cause was by chance actually observed. When the wind was strong and in a certain direction, it blew through the outside grille and through the hatch into the ante-room. If the connecting door was closed but not properly latched, it blew open; being sprung, it then closed with sufficient force to latch it. Now shut and latched, it attracted no suspicion. The door catch was found to be poorly fitting so that it needed some pressure to latch, and consequently was frequently left unlatched. Attention to the latch eliminated the false alarms.

Catches, latches and locks on wired doors and windows should be checked at the time of installation, but they could deteriorate afterwards. Alterations such as fitting draught excluders may be made, which could cause a poor fit and the possibility of a door springing open. All those on the affected zone should therefore be examined. Users tend to be careful with exterior doors, which of course they should, but are often less so with interior doors.

False alarms and PIRs

The PIR detector is normally reliable and not prone to generating false alarms as are ultrasonic and microwave devices. This is because it responds only to objects that both move and radiate infra-red. If properly sited there should be no trouble, but some possible alarm sources might have been overlooked. A grandfather clock having a pendulum that reflects the sun's rays from a window at a certain time of day is an example.

Usually heating systems cause no trouble, except forced air heating, but it is possible that a strong draught of warm air getting into a cold room from a heated part of the premises could be seen by the sensor as a moving warm object.

Where there are resident mousers or other pets, a false alarm is possible unless the sensor is carefully sited to avoid them. An animal could have been introduced into the premises after the system was installed, or a large rodent could be responsible.

Where two or more PIRs are connected to the same zone, they should be latching, so the one causing the alarm can be identified. If this is not the case, false alarms can be much more difficult to trace.

If there is no stimulus the PIR cannot trigger, even if set to highest sensitivity, but if there has been an inexplicable false alarm, sensitivity can be checked by a walk test to see that it is just sufficient to trigger with a human moving at the limit of the protected area.

Wind and storm

Stormy conditions often produce a spate of false alarms, but do not too readily assume that they are false. Burglars frequently choose windy nights for their activities because any noise they may make is masked by that of the gale.

There are several ways whereby a storm could actuate an alarm system. Inertia and vibration sensors are perhaps the most vulnerable as vibrations and movements of building fabrics and fencing are likely to considerably exceed the normal. There is little that can be done to prevent this as reducing sensitivity will only make them too insensitive to adequately protect against intrusion. If a violent storm has been forecast and all the inertia sensors are on one zone, that zone could be left off while the rest of the system is on. This of course would mean some loss of security, but it would have to be balanced against the trouble and inconvenience caused by a false alarm. Much would also depend on the degree of back-up cover given by the other types of sensor.

Exterior windows and doors fitted with magnetic switches should not be affected by wind unless actually blown open. They should be physically secure enough to withstand rattling that is of sufficient violence to set off the alarm. If not, then the physical security needs improving. However, interior doors with poor latches could be blown open, as the example described earlier illustrates. All interior doors that are wired should have positive latching action. When investigating, do not overlook the possibility that the door might have been closed or blown shut after initiating the alarm.

Water could cause an alarm if it saturated a sensor terminal block or terminal box as it could cause a leakage between the detector and anti-tamper circuits if fitted. This is more likely if it contains dissolved salts leached from building materials which increase the conductivity of water. Sensors and junction boxes should be checked and dried out wherever there is any sign of water entering the building. Domestic systems lacking an anti-tamper circuit should be unaffected.

Wiring faults

If greatest care has been taken with the connecting of detection loops as stressed in the previous chapter, there should be no reason to suspect bad connections. However, the system might have been installed by someone else who was not so meticulous.

Another possible cause is work by builders, decorators, electricians, plumbers or just someone hammering a nail in a wall to hang a picture. Cables can be and have been damaged in such a way as to cause intermittent faults. This may not show up immediately, so think back as to any work carried out in the premises for at least the previous month.

Other changes such as furniture rearrangement, installation of new equipment, clearing out, and even spring cleaning, are all possible causes of cable damage or disturbance to sensors, especially space sensors, and should be considered.

Misaligned magnetic switches

This is not the most likely cause of trouble, but is easiest to test for, and so is worth trying before embarking on any dismantling. Misalignment is really due to bad installation, but the trouble may not surface for some while, so its true nature may not be suspected. If a switch and its magnet are not aligned when the door is closed, the magnetic field acting on the reed is smaller than it should be. It may be just sufficient to hold the reed switch on, but there is no tolerance for further field reduction.

All magnets slowly lose some of their magnetism. Also, the gap between frame and door can increase owing to wood shrinkage, or the door may drop or warp, thereby reducing the magnet's influence. So, the switch may function for a while, but its operation may become increasingly critical, sometimes closing and other times not. The result can be perplexing false alarms.

Not to be overlooked is the possibility that the switch and magnet were properly installed originally, but later a new door was fitted, and the magnet was then incorrectly aligned by the fitter. This has been known.

Any suspected magnet/switch combination can be easily tested by actual operation, using the control panel internal buzzer. It should be possible to open a door at least an inch before the alarm triggers. If it is actuated by a much smaller door movement, then the set-up is too critical and is very likely to cause false alarms. If the alignment appears right and the door gap is not excessive, the magnet may be weak or the reed too insensitive to magnetic fields. Replacements will prove which.

Cable breaks

Breaks in the closed sensor loop can be difficult to find without expensive test equipment and the knowledge of how to use it. However, the following method which requires only a simple continuity tester will eventually run the fault to earth. First, though, make a careful examination of all sensors and wires leading from them, and all exposed loop wiring, to see if there is any obvious damage. Failing this, proceed as follows.

Disconnect the loop at the control box. Choose a sensor half-way around the loop and remove it to gain access to its terminals. Connect the tester from the brass anchoring terminal to one of the silver switch terminals. If it reads OK, the furthest half of the loop is in order and the break must lie between that sensor and the control unit. If it is open-circuit, the fault is in the furthest part.

Dismantle another sensor in the faulty half and repeat the test to see if the fault is between there and the previous test point. Thus the defective section is narrowed down until the actual fault is found. It can be a rather laborious process and chance plays a big part: you may hit the fault first time or it may stay hidden until the last.

Intermittent faults take a lot more patience. If the loop cable seems undamaged after a careful visual inspection, the fault can be in either the connections to a sensor or the sensor itself. If, though, window foil is included in the loop, this should be the first suspect, as the contact between foil and terminals can deteriorate or the foil itself develop a hairline break. The most practical way of dealing with the problem is to just short-circuit the window foil for a period and see if any false alarms are then experienced. If not, the foil is most likely to blame and should be completely replaced.

If the fault persists, before doing anything drastic to the loop, it would be as well to prove that it is not a fault in the control unit or other sensors. To do this, disconnect the closed loop and short-circuit the terminals. Run the system as normal for a while. Back-up sensors such as PIRs and pressure mats will give a measure of security for the period that this test is in progress.

If after a reasonable period no false alarms have occurred, the control unit and the other sensors are absolved from blame and the fault must lie in the loop. The next step is to dismantle every sensor and remake the connections. The fault may well then be found to have cleared and no further trouble be experienced. If it persists, then it is best to replace every magnetic switch. This is known as a 'block' change and it means that many good switches are discarded along with the faulty one. Although this seems wasteful, it is better than having persistent false alarms. It is, though, unlikely that you will have to go this far; most of the loop faults are due to damaged wiring or poor connections.

Control panel and power supply faults

These are the least likely to cause false alarms. By far the majority of faults are those already discussed. However, they do sometimes misbehave. Testing is quite easy. Remove all detection and anti-tamper wiring for the affected zone and bridge their terminals. Silence the external sounders and disconnect any telephone connections. Then switch on and operate the appropriate zone switch.

Movement detectors can be triggered by a change of supply voltage, which in turn can be produced by mains voltage variation or a change in load due to some other device on the auxiliary supply circuit. This will not happen if the circuit voltage is regulated, which it should be in any reputable control unit. For technical readers, a test of its regulation can be made by measuring the voltage, then shunting a 50 Ω load resistor across it and re-measuring. The difference between readings should be very small.

After a false alarm

Some authorities recommend that the system not be switched on again after an unexplained false alarm until a thorough test can be made of the system. Frankly, the value of this advice is very dubious. Some burglars are known to deliberately set off an alarm system without obvious entry, then wait for the alarm to be switched off and the fuss to die down before breaking in without risk of interruption.

How could an alarm be set off externally? One possible way is by using a magnet. Remember that the reed switches are held on by the magnets in the doors or windows; they are released by removal of the magnetic field. If a bar magnet producing a similar field is held against the door parallel with the one inside it, but with magnetic polarity reversed, it could cancel or divert its field sufficiently to release the reed switch.

The position of the sensor can be easily determined by running a pocket compass around the frame to detect the presence of the door magnet. The magnet strengths can then be matched if the applied magnet is stronger by gradually moving it away, turning it to achieve polarity cancellation. Some skill and patience may be required but it is possible. A more likely method is a hefty shoulder charge to momentarily displace the door magnet.

Rather than switching the entire system off after an apparently false alarm, there are several preferred options. The simplest, if the system is multi-zone, is to cancel the zone that originated the alarm, but leave all the others on. This provides some security at least, and intruders are likely to trip sensors on another zone if they break in.

If a telephone dialler is connected, this certainly should be disconnected to avoid raising a false alarm at a distance with possible police involvement. Another possibility to avoid disturbing nearby neighbours would be to disconnect the external bell but leave the internal connected. This would still serve as a powerful deterrent, yet would cause minimal disturbance to the neighbourhood.

Maintenance

Regular maintenance is advisable for all security systems, whether a simple owner's check of a domestic system, or a professional engineer's test of a large industrial installation. In the latter case, maintenance is essential.

For domestic or small shop premises, the owner will be aware of any accident or damage that could affect the system and will take steps to rectify it as soon as possible. However, a regular routine check should also be made.

The outside bell is the most vulnerable of all parts of the system, not only to interference but also to the effects of air pollution. Acidity in industrial areas and salt near the coast can cause corrosion, while other pollutants can gum up the works if allowed to accumulate. It is good practice, then, to sound the bell by triggering the system once a week.

This will confirm that the bell is working, and keep the moving parts free. If this is done on the same day and at the same time each week, and allowed to run for only some ten seconds, neighbours will not confuse it with the real thing, and it should not create a nuisance.

It is said that most opportunist break-ins are by locals. A regular sounding of the bell will tell the neighbourhood that you have a working alarm system, and the box, If used, is not just a dummy.

If a different sensor is used to trigger the system each time, this will test them all in the space of a few weeks. If standby dry batteries are used, a test every two months with the mains supply off will ensure that they are serviceable. Contrary to what may be expected, this will not deplete them if the test is only for a short time. Dry batteries deteriorate most quickly when they are not used at all; small periodic discharges actually extend their life. Given small regular discharges, the life of the batteries should extend more than two years, but to be sure it is probably best to change them then.

11 Gadgets and gimmicks

In addition to the conventional alarm systems and regular physical protection devices, there are a large number of security items that are now available from mail order and other suppliers. Some of these are very good, many are helpful, but others are of dubious value.

Heading the list of the first category must be the PIR floodlight. It has been described already in previous chapters, but no apologies are offered for mentioning it again. It really is an excellent deterrent and one of the few devices that offer maximum protection with no inconvenience; in fact it can even make life easier.

For readers who might have skipped the references in earlier chapters, the PIR floodlight is a device that is actuated by a moving source of infrared radiation, which includes the human body. Its range of detection spreads out fanwise for up to 40 ft (12 m). When anyone moves within range, the floodlights come on and stay on for a pre-selected time. When that expires they go out, but they immediately come on again for a further similar period if the body is still moving within range. As they thus do not go out permanently until the area is clear, a short time can be pre-set, usually 15 to 30 seconds. This prevents wastage of electricity when the intruder is no longer in range.

Daylight detectors are included which automatically switch the device off during daylight hours and on again when darkness falls. So no attention is needed at all, other than to replace lamps at very infrequent intervals. The only installation required is a mains supply and the actual mounting of the self-contained unit on the outside wall.

If mounted high on the rear wall of the house, it will effectively protect all its usual entrance points from intrusion at night, as few if any burglars would be prepared to attempt a break-in while fully floodlit! Further protection at the front could be obtained by another device mounted there, and this is recommended especially for remote houses or those with concealed frontages. Dark side alleys could also be made safer with detached, semi-detached or end-of-terrace houses by a PIR floodlight on the side of the house. Prices have tumbled in the last few years, so there is no limit to the number that could be installed.

A plus feature is that legitimate visitors will be illuminated and so able to negotiate steps or other hazards as soon as they approach. Also it provides lighting for the occupants if they need to go into the garden or out the front after dark, without having to switch porch or other conventional lights on and off.

Just one word of caution, though: do not place total reliance on them. Remember that they can have no effect on a daylight break-in. So good physical security along with an alarm system is still the essential protection formula. PIR floodlights and any other single device should be used only to complement, back up, or increase security obtained by the standard methods.

Self-contained alarms

These consist of a PIR or ultrasonic detector, a sounder which is usually an oscillator and small loudspeaker, and a dry battery power supply. The whole lot is contained in a small case which is often disguised to look like something innocuous such as a book. They are frequently sold as complete installation-free alarm systems that need no wiring and as such would seem to offer advantages over the conventional installation.

They do, though, come in the 'dubious' category. As we have seen, the first line of defence must always be the perimeter. It must be made physically secure and protected by alarm sensors that will sound as soon as the perimeter is disturbed or breached, so keeping the intruder out. Internal space detectors should be used only to back up the perimeter defences. Self-contained alarms allow the intruder to gain access into the protected room before any alarm is sounded. Furthermore, the PIR devices can only protect one room, so several would be needed to protect a whole house.

One type of self-contained alarm, which is claimed to protect the whole house, is incorrectly described as ultrasonic. Ultrasonic detectors work by generating an ultrasonic sound and receiving it back from moving objects as described in Chapter 6; they can only protect the room they are in. The devices described here work by a change in air pressure. If a window or door is opened anywhere in the house, or a window broken, the air pressure changes and so actuates the alarm.

As it does not respond to movement, the device can be on when the occupants are at home and engaging in normal activities. This is an advantage, as it can protect the home from entry in one place, such as the back, while the family are elsewhere such as in the living room watching TV.

All internal doors would have to be left open or ajar so that pressure changes could readily be communicated to the device. If one was left closed, that particular room would be isolated. The possibility of this happening, especially in the winter when one gets into the habit of shutting doors, is the weak point of the system.

Another weak point which it shares with all self-contained systems is that of its internal sounder. Although some devices such as the book look-alikes are disguised, the direction of the sound gives the source away, and enables an intruder to quickly find it and deal with it. He could silence it by pushing it under a carpet, piling cushions on it, or smashing it: few would continue to sound after being jumped on!

The sounder emits a shrill tone which is claimed for most models to be in excess of 100 dB. However, the makers do not stipulate the distance at which this is measured. As shown in Chapter 7, sound pressure levels decrease by 6 dB for each doubling of distance. The standard measuring distance for alarm sounders is 3 m; but some are quoted for 1 m, which is 10 dB more and looks a lot louder on paper than it really is. There is nothing to stop a manufacturer quoting the level at a few centimetres, which could well be over 100 dB, but the level would be a lot less at a distance of several metres.

The acoustic sound power, which is related to sound pressure and the volume we hear, depends on the electrical power that is fuelling the source. You cannot get more power out than you put in; actually you get a lot less because no power conversion from electrical to sound can be 100 per cent efficient. So, if the device is powered by small dry batteries, the sound output cannot be anywhere near as great as that from an alarm system bell or siren which is powered either by large batteries or by the mains. Furthermore, as the sounder is inside the premises it will not be heard very far outside so as to raise an alarm. At best it would startle the intruder, and possibly alert others who may be in the house.

The air-pressure device in particular is a promising idea, but would be better used in conjunction with an outside sounder and power supply, in other words as part of a conventional system. The advantage would be in the need for just one or perhaps two sensors to cover the whole house, so greatly simplifying the installation.

All this is not to say that self-contained alarm units have no use at all. In flats or bed-sits they can indeed provide a measure of protection for individual rooms and alert occupants of adjoining accommodation that an intrusion is taking place. In this situation it is as well to tell neighbours about the unit and let them hear it, so that they will recognize it if they hear it again. One such unit could also be used for holiday flats, where burglaries are often reported, and even hotel rooms.

So, they are of value in those limited circumstances, but should not take the place of a conventional security system in a normal house.

Personal attack alarms

Much of what has been said above applies to personal attack alarms, so we will briefly mention them. The situation requiring them is quite different. While they suffer from the same limitations, you cannot carry around a full-blown alarm system, so there is no other option, especially as the law seems to frown on pre-emptive self-defence. Whether or not you choose to carry something to use against an attacker and thereby risk prosecution, or submit to being injured, raped or killed yourself, is a personal choice which is forced upon us by the law as it is at present. Many feel that Mr Bumble's famous words certainly apply here: 'the law is a ass – a idiot.'

A personal attack alarm may indeed succeed in scaring off some attackers; even if snatched and thrown away, it could alert anyone within earshot that something is amiss. But no one, especially women, should rely on such devices. Rather, people should take sensible precautions such as avoiding dangerous and lonely areas when alone, keeping a good look-out when walking, being ready to run if necessary, locking the car when stationary, and other measures which may be prudent for the circumstances. The alarm, though carried, may then never need to be used in anger.

Door-viewers

The door-viewer is a tiny lens that is fitted into a hole in the front door at the occupier's eye height. It gives a panoramic view of the area outside. By this means the householder can see a caller and identify him. You cannot see in from the outside, nor can you see if anyone is looking out. So if the occupant does not wish to answer the door after viewing, the caller has no means of knowing that he has been observed or even that anyone is at home. The only time the observer may give himself or herself away is if he or she puts a light on when it is dark, as light can be seen from the outside.

It is a cheap, effective, and easily installed device, and is recommended especially for those living alone or the elderly. It is less effective if there are glass panels in the door as, even if they are frosted or semi-opaque, anyone close to the door can be seen from the outside.

Door chains

These are another simple, easily fitted and largely effective means of protection. If the door is opened while the chain is engaged it cannot be forced open any further, and it actually needs to be drawn together before the chain can be released. Thus undesirable callers cannot force their way in, and a limited conversation can be held through the partly opened door before the occupant decides to admit the caller or otherwise.

An important feature to look for is that the joins in the links of the chain are welded, not just pressed together. If this is not so, a hefty shove against the door could force a link apart and break the chain. When fitting, long screws of adequate gauge must be screwed into solid wood. If the door frame is weak or the screws are too short, the fitting could be burst off. Usually, screws are supplied with the chain.

There is just one criticism. It is true that if the chain is too long, it could be a security hazard, as the intruder could reach in and disengage it, but most seem to go to the other extreme of being too short. In some cases where the door frame is hard up against an abutting wall, the door cannot be opened sufficiently to see around it, which means that the chains are then unusable. There really should be two sizes: normal, and a slightly longer one for such situations.

Door wedge alarm

This device is a wedge-shaped object with a circular housing at its thick end that contains a battery-powered mini-siren. It is placed so that the thin end is wedged under the front door. If an attempt is made to force the door open, the device may jam it and prevent further door movement. At the same time the siren is actuated.

This is one of those gimmicky devices that are of very dubious value. For a start, it can only be used when the householder is at home as he obviously cannot leave with the wedge in place. Also there are many doors that have draught excluders which would simply push the device away when opened. As we have already seen, the front door, if substantial, is the least likely entry point to be attacked; it is the windows, especially at the rear, that are most vulnerable. If the front door is weak, it would be far better to spend money on getting a stronger one than on one of these. If there is a good alarm system installed they are unnecessary; if there is not, they are certainly no substitute.

Door entry systems

Remote door entry devices are invaluable to the elderly, the infirm, flat dwellers and others. The system comprises a loudspeaker/microphone unit that is fitted to the door, a telephone to which it is connected inside the premises, a door-lock release and a power unit.

The speaker and telephone units allow the occupant to converse with the caller. If satisfied as to his identity, the occupant presses a button and the lock release disengages the door catch, allowing entry.

The lock release is fitted to the door frame in place of the normal staple into which a springlatch or deadlatch engages. The existing lock thus continues in use. As the caller must pull the door shut to secure it when leaving, the system cannot be used with a security deadlock, but it can be used with a deadlatch.

Two separate power supplies, one DC for the speaker phone and the other AC for operating the release, are contained in the power unit. Five wires are needed, so six-core cable must be used.

Two or more phones can be connected to the one door unit, and multiple systems are commonly found in high-rise flats. The caller selects the required flat number by entering it on a key-pad which is part of the speaker unit. Some of these systems have a timeswitch which releases the lock when a tradesman's button is pressed during restricted periods. At all other times the button is inoperative.

Intercom door chimes

This is a radio system. The transmitter, with press buttons for calling, is fitted to the front of the door or a nearby position. The receiver is hand-held but has a belt or pocket clip. Both contain loudspeakers that double as microphones.

When the button on the transmitter is pressed, a chime is sounded in the receiver, whereupon a two-way conversation can be held with the caller. No wiring is required, and the range is around 150 ft (45 m), so that it can be operated from anywhere in a large house, even the bathroom, and from the garden. It enables the occupier to move around at will without risk of missing important callers, while filtering out the undesirable ones.

The transmitter is only activated and consuming power when someone calls, so it can be run on four small torch batteries. The receiver must be on standby all the time, which consumes current. It therefore is run from rechargeable batteries, and a charger is included in the outfit.

The system provides security by identifying callers without even having to go to the door, and undoubtedly is easy to install and convenient to use, unlike many security devices. The batteries must be kept charged, but the receivers have 'low battery' indicator to warn when recharging is required.

The only reservation is whether the transmitter would stand up to vandalism or even theft. While crime is now virtually universal, vandalism does seem to be worse in some areas than others. So, if the area has a bad record for vandalism the transmitter could attract undesirable attention and not last very long; if not, and especially if you have a large house or are infirm, it could be a very good buy. The unit described is sold by the mail order firms Innovations and Direct Choice of Swindon; the latter do not charge for packing and postage.

Switching lights

A cheap and popular deterrent is the device that plugs into a lampholder and into which the bulb is inserted. The light will come on for a certain period and then switch off and come on again later. A number of these operating in different rooms when the occupants are away will give the impression that the house is occupied.

They may not deceive the more canny burglar, especially if he sees lights going on and off in different rooms in the middle of the night when most people are in bed and asleep, or if they come on during broad daylight. He knows about them, so they could actually inform him that the house is unoccupied. However, they are likely to deter many potential intruders and so on balance can be recommended.

Some have daylight inhibitors that prevent them coming on when it is light, and these are certainly to be preferred – for electricity saving if for no other reason. Others do not have the intermittent switching facility, but merely switch on when it is dark and off when it gets light. Being on all night, these give a clear indication that no one is at home and so are of very dubious worth. Another variation is the listening light which comes on when it hears a sound. A good place for these is in the bedroom, where it is natural to expect a light to come on if the occupant is disturbed by a noise.

A light should never be left on in the hall when the occupiers are absent. Few people leave a light on in the hall when they are at home, unless inadvertently. Yet many do so when they go out, either to light them when they return, or to make it appear (as they think) that someone is at home. Burglars are quite aware of this quirk, so a light on in the hall is a dead give-away.

Beware of the dog!

We can hardly classify man's best friend as a gadget or a gimmick, but we have to put him in somewhere! A dog is undoubtedly a deterrent, especially a large one, though often the small ones bark the most. However, it would be unwise, as some do, to rely solely on a dog for protection. When owners are on holiday, which is when most break-ins occur, the dog is usually absent too.

Many break-ins have been carried out in spite of a dog. Doped meat pushed through the letter box is one ploy that has been used. A legitimate caller once visited a house when the owners were not at home. A dog came to the front door and barked ferociously. The caller opened the letter-box and barked back – and then saw the dog scurry into the kitchen and hide under the table!

So, as with most other items of ancillary 'equipment', a dog can be a useful security booster as long as it is not relied on totally. Another factor to be considered is possible false alarms due to the animal stepping on pressure mats or actuating space sensors.

Bark on record

There is a device that plays a recording of a ferocious-sounding dog barking whenever the bell push is operated, and continues for a minute or so. Cheaper than actually keeping and feeding a large animal, it seems that it must be a good idea. There is just one snag. There is often a pause of a few seconds while the tape starts up and the playback begins. This of course is quite unnatural: dogs start to bark immediately the bell starts sounding. A delayed-action dog is rather a give-away. Also, there is no barking if the knocker is used instead of the bell. So this doggy device is rather a dodgy device.

An improvement would be for the device to be initiated by a microphone so that it would respond to any loud sound near the front door, and also of course if it could be made to start instantly.

Wireless alarm system

This is not so much a gadget as a complete security system, but it has been included in this chapter because it is out of the ordinary run of security systems. The model here described is the SK 8000 made by Celtel of Basingstoke, Hampshire.

The kit comprises a control box, three magnetic switches, a PIR detector, a power horn sounder, a flashing lamp module, two remote control setting units and ten batteries.

The control unit contains a loud horn sounder and will accept up to sixteen sensors; indeed the three that are supplied would not be adequate for most installations. Each sensor has a separate transmitter box to which it is wired, and which must be mounted nearby. Normally the transmitter is inactive, but it transmits a short signal every 1A hours to inform the control box it is functional. If this signal is not received, a fault is indicated on the control panel against the number describing the particular sensor. If the sensor is actuated, the transmitter sends a pulsed signal which, when received by the control box, sounds its internal horn and also the separate power horn.

To enable the control box to distinguish between the different sensors, the signal from each has an individual pre-set unit code. In addition, there is a house code common to all sensors on a particular system which distinguishes it and so prevents interference with any other system nearby. The operating transmitter frequency is 1.418 MHz.

The power horn supplements the sound from the internal control box horn. It gives a claimed 110 dB, but the measuring distance is not specified. However, as it is mains powered, it could indeed produce this at either of the standard distances of 1 or 3 metres. The makers suggest it for the bedroom, but this is questionable. The maximum time that 111 dB is permitted in the workplace is 3.75 minutes, otherwise permanent hearing damage can occur. This time could easily be approached during waking, getting up, and taking action to silence the horn.

There is no provision for an outside sounder such as a bell, but the power horn could be mounted outside if enclosed in a weatherproof box. Although it is radio controlled, it needs a mains supply, which could be more of a problem than just running bell wiring. There is also the point raised in an earlier chapter about the possible confusion that could result from the use of horns for home security systems.

A big plus feature with this system is the complete avoidance of the exit/entry problem. It is set by means of a key-fob wireless remote control when outside the house, and is turned off again by the same means. Although only one fob is supplied, up to eight can be used with any one system, so individual family members can each have his or her own. The fob also serves as a panic button, but the snag with this over conventional buttons is that one may not have it handy when needed. Installed buttons are always there. The key-fob is also house coded so that other fobs are unlikely to control the system.

There is no provision for pressure mats, which is rather unfortunate as these are a cheap, uncomplicated and valuable second line of defence and are especially useful for certain situations, as we have seen.

What then are the pros and cons of this system? It is certainly easy to install, and is especially useful in large rambling premises providing a sufficient number of extra sensors are fitted. The remote setting and unsetting is a big advantage over conventional systems. As we have observed though it does have negative features, to which must be added its initial cost, and the recurring cost of replacing batteries in each sensor – although these do last up to a year.

The sensor transmitters are rather cumbersome. They have test and delay controls which need attention, so the units must be mounted in an accessible position. A wall-mounting job for each thus appears to be inevitable. With modern micro-transmitter technology, one would have expected a rather more compact unit.

Conclusion

We have considered some representative devices currently available. It is hoped that these critical appraisals will have shown the reader the sort of things to look for and the sort of questions to ask about any new wonder gadget that may yet appear. New products are appearing all the time, so there will be no shortage of subjects to investigate!

The general conclusion, though, is that the conventional wired perimeter intruder alarm remains the backbone of the home security system, in conjunction with good physical security. Other devices can be useful back-ups or supplements, but can never take the place of the established methods.

12 Watch out!

There is little point in conscientiously attending to every detail of physical security and installing an effective alarm system, if you then open the door to an intruder and let him in! Or if you advertise to all and sundry that you are away on holiday; or if you sometimes neglect to properly secure the premises. As the ubiquitous poster warns: 'Watch out, there's a thief about.' You cannot relax vigilance for a moment.

This may sound depressing, presenting a picture of always looking over your shoulder and thinking of nothing else than possible attacks on yourself or your property. Although the present increase in crime has created that sort of situation, constant vigilance is not quite so bad as it may at first seem. It involves firstly thinking of all possible situations that could spell danger, and then developing a routine of what to do and what not to do in given circumstances, and always carrying it out. This will soon become automatic and not seem at all onerous. We will look at some of the common pitfalls.

Lock up – always!

Many people make the mistake of slipping out for a few minutes, perhaps to a local shop, and leaving a window, the back door, or even the front door, unlocked. As stated before, most burglaries are done by locals, usually youths looking for an opportunity. If they happen to see you leave, and know that no one else is likely to be at home, they will be in and out in a matter of seconds. They may not take many items, but the one or two they do take could be of value and potentially expensive. Wallets, purses, handbags with money and credit cards are a prime target for these quick in-and-out thieves.

The only way to defeat them is never to leave the premises with any ground-floor window open, or door unlocked, back or front. Yes, switch the alarm on even for that quick pop-out. If it has been designed to give 'friendly security' as recommended, this will not be a chore, and it is more likely to be used for every absence no matter how short.

Another open invitation to thieves is when unloading the car. This often entails carrying in parcels with both hands full, thereby leaving the hatch or lid of the boot wide open. You may be inside for only a matter of seconds before coming out for the rest of the load, but a chance passer-by could lift an item and quickly disappear in that time. It has often happened.

So, when unloading the car, always shut it even if it means putting parcels down on a wet pavement to do so. It may not be necessary to lock the boot as there is nothing on view, but at least shut it. If you are likely to be indoors for more than a few minutes, it would be prudent to lock it as well. Getting into the habit of always doing this will make it seem less inconvenient.

Although this is not strictly guarding your home, it must be said that many if not all car thefts by youngsters, and therefore many deaths caused by 'joyriding', could be prevented. Amazing as it may seem there are still car owners who not only leave their cars unlocked, but leave the keys in the ignition. There are many ways a car can be immobilized; one of these should be used, and the car should always be locked, even for short periods. Again, habit will make this precaution automatic, and so possibly save the life of an innocent 'joyriding' victim.

Holiday precautions

As mentioned before, holidays are a prime time for burglaries. The thieves have plenty of time to enter and search, and can come back again to remove further valuables if they wish. Houses have been completely stripped while their owners were away on holiday.

In addition to securing the house, every effort should be made to avoid making one's absence evident. The obvious things are to cancel papers, milk and any other regular deliveries. But when cancelling the papers, do so by phone or, if in person, ensure no one else is in the shop while giving your address and times of absence. That man browsing at the magazine rack could be listening with keen and nefarious interest!

Neighbours should be informed if you are sure they are trustworthy. Not all are. Burglars have to live next door to someone, and it could be you; but, to be sure, they are not going to advertise the fact. If you feel happy about them, get them to keep an eye on your house, and be alert for any unusual activity and strange callers. Get them to remove junk mail from your letter-box and mow the lawn if they are willing. Of course you must be prepared to do the same for them.

You may feel confident enough to let them have a key to open the front door and collect up any mail that could be visible from outside through glass panels or the letter-box. This could also foil a common burglar's ploy. Some thieves, if they suspect a house is unoccupied, set a trap in the form of a leaf or some other innocuous fragment which they wedge between the front door and the frame. When anyone enters, the leaf falls out unnoticed. If it is still there after a couple of days, the crooks know that no one has been in, and so the owners are away.

For neighbours to be able to enter, the front door must not be on the alarm exit circuit. It can have a lock-switch to bypass its sensor, or it can be left off altogether as we have seen in earlier chapters. The latter has the advantage that the system will still be on when the neighbours enter. If they are warned about this it will prevent possible prying as they can proceed no further without actuating a sensor and setting off the alarm. Even if they are trustworthy it is wise not to acquaint anyone outside the family of the details of the alarm system.

The police could be informed. Although they may not have the manpower to keep a special eye on your property, they will know that if there is a report of an alarm, it is unlikely to be a false one.

If you have reason not to entrust the neighbours with your keys and you have a glass panelled front door, cover the bottom panel with cardboard or even paper, so that the growing pile of mail inside is not visible.

Another ploy which has frequently been worked by the criminal fraternity is to look out for baggage at the airports bearing the owner's name and home address, note it, then pass it to criminal colleagues. If any luggage identification is deemed desirable at all, just inscribe your name and home town. To do otherwise is to let the world know that the occupants of your address will be a long way from home for a long time.

Secure garden tools

The traditional picture of the burglar as a man with a mask, striped jersey, a sack labelled 'swag', and a jemmy, is of course far from reality, although he may at times carry the long metal bar with flattened end for levering and breaking known as a jemmy. Being found in possession of house-breaking implements could lead to instant arrest if apprehended, so he usually makes do with something less obvious, such as a large screwdriver and a few other odds and ends.

Often, though, he doesn't need to carry such items around because he can find suitable ones on the spot. A garden spade, hoe, and other such tools will serve for all sorts of levering and breaking purposes. So don't provide him with the means to break into your home; keep them locked away and out of sight and, if they are stored in a garden shed, make sure the lock and fastenings are secure as described in Chapter 1.

A certain way to help the burglar is to store a ladder where it can be removed and used to reach your upper storeys. Apart from this, the ladder itself is likely to disappear, so if you have one, always keep it chained up with a solid chain and padlock.

Tricks and treats

Some crooks use cunning instead of brawn. A few examples of actual cases illustrate the sort of thing to beware of.

A street busker went into a pub and asked if he could leave his violin somewhere handy for him to collect later, as he had to see someone and didn't want to take it with him. The landlord agreed and put the instrument on a shelf behind the bar. A day or two later a prosperous-looking man came in and, after ordering a drink, caught sight of the violin. He said he was a dealer and asked to see it.

On examining the instrument he asked if it was the landlord's property. When told it was not, he said he would pay £5000 to the owner for it. The landlord, sensing a good opportunity, asked him to come back and he would try to negotiate a deal. When the busker returned the landlord offered to buy the violin for £100, but the man refused, saying it was his only means of livelihood. The offers were increased until the man eventually agreed to accept a £1000 cheque written to bearer. The 'dealer' never returned and, when valued, the instrument was found to be worth less than £50. The cheque had been cashed the same day.

In another case, a car was stolen, but was returned the next day with a note of apology explaining that the writer's own car had broken down and he needed to get to an urgent appointment. To compensate for the inconvenience he enclosed two theatre tickets. The owner and his wife were very pleased at this, and went to the theatre at the appointed time, looking forward to a pleasant evening. It was dashed on their return when they found that their house had been almost cleared.

These tricks succeeded because of greed, or the overwhelming desire to get something for nothing. It is the basis for most confidence tricks, as they are called. So if you are offered something which seems too good to be true – it probably is!

Not all tricks have this basis, though; some are opportunist. An example is the case of a couple who hired a taxi to go to the theatre because they did not want to take their own car and have the problem of parking. On arriving and after dismissing the taxi, they found to their dismay that the tickets had been left at home. It took a while to get another taxi, but when they eventually did, they returned home to find their house being ransacked – by the first taxi driver.

In another trick which has been performed many times, the victim receives a phone call from someone claiming to be the police, with the message that a family member has had an accident and is in a certain hospital. The victim anxiously leaves in a hurry, neglecting to fully secure the house. So, the crook is not only presented with an empty house for a good while, but one likely to offer easy entry. By the time the victim has gone to the hospital, made extensive enquiries, and finally returned home perplexed, the burglars have taken their pick. The lesson with this one is that if you receive a message which has the effect of getting you out of the house in a hurry, always phone back to the supposed caller to verify it.

Another, more elaborate trick using the phone was once played on the matron of a private nursing home. Having just drawn money for the staff wages from the bank, she received a phone call seemingly from the bank telling her that they had reason to believe that some of the notes they had given her were forged, and they were sending someone around straight away to examine them. In the background she could hear familiar bank noises and had no reason to suspect the genuineness of the call.

The visitor, a pleasant well-dressed young man, examined the notes and declared most of them to be forged although some were genuine. He took the 'forged' ones, gave her a receipt on the bank's headed notepaper and said he would return with the equivalent amount. That was the last she saw of him or the money. The bank had not phoned; the background noises were probably recorded in the bank and played on a recorder; the paper was a photocopy of a genuine letterhead. Leaving some 'genuine' notes was the master stroke to allay any doubts and establish confidence. Again, the loss would have been prevented had the matron called the bank to check, immediately after the bogus call.

It is easy to be wise after the event, and most people would probably have fallen for any of these tricks. Others have been devised and no doubt there will be many more. So the golden rules are: if you are offered something for nothing, an unbelievable bargain, or are presented with any situation which promises riches, be highly suspicious. To reverse the well-known proverb: every silver lining has a cloud. Check out any unexpected call which would get you out of the way if you followed it.

Gabbers and seekers

These are not pop groups but the underworld trade name for tricksters who call at the home in what the police classify as 'burglary artifice'. This type of crime is growing at twice the rate of others as more houses are made secure against forcible entry. Vigilance is needed lest all the efforts to guard your home from break-ins are nullified.

The tricksters work in twos. One is the front man with a plausible story to distract attention; he is the 'gabber'. The other is the 'seeker', the one who nips in unnoticed, or poses as an assistant to the first and who does the stealing. The guises used are almost endless. Among them are the following:

1. Plumbers who have mended a leak in the upstairs flat, and who have called to check for contamination. While one is making 'tests', the other is removing what valuables he can find.
2. Council surveyors who come to check the flats. One gets the victim to show him round, while the other makes a few spot checks – on movable property.
3. A distraught man comes searching for his dog; could he look around the garden? As the victim helps him look, the seeker slips in the open door.
4. Similar, but children looking for their ball. The victim accompanies them to make sure they do no damage while a youthful seeker whips purse or handbag from inside.
5. A child comes asking for a drink on a hot day. Who can refuse? While the occupier is getting something, and/or while the child is drinking it, another slips in the open door.
6. Water company officials come tracing leaking pipes.
7. Social workers visit to check the welfare of young children. These are especially to be guarded against, as abduction rather than theft may be their aim.
8. Welfare officers come to check if the victim is getting the maximum benefits he or she is entitled to. A clever one this, as it dangles the prospect of more money, which ensures full cooperation.
9. Council tax inspectors come to give a refund – another crafty one – and offer a large-denomination note, asking for change. Victims thereby reveal where they keep their cash.
10. An antique dealer offers a huge sum for something he sees just inside the front door of a large older house. He thus gains entry to see what else there is and whether it is worth a burglary later. Needless to say he never returns to buy the article.
11. A motorist has just broken down outside: can he make a phone call to his garage, and one to his wife who is expecting him home? The door is left ajar while he does so and the victim hears him talking loudly, but doesn't see the seeker slip in. A generous sum left to cover the calls allays suspicion – until items are later found to be missing.
12. A builder calls to say he has noticed a fault in the roof: can he point it out? There may be some minor fault to justify his claim, for few roofs are perfect; but while the victim is looking skywards and listening to the gabber's spiel, the seeker is inside attending to more mundane matters.

All of course are bogus, and these are just a sample of the various ploys that are used. How then can one determine which are genuine callers and which are not? What defence is there?

Firstly, do not let your guard down: be suspicious of all callers unless you know them, or are expecting them. Be especially suspicious of pleasant smooth-talking individuals, because that is the manner cultivated by the typical gabber.

Do not be afraid to ask for identification. All reputable callers will have an official ID card, usually with a photograph, and will normally volunteer it without asking. However, anyone could have a passport photograph taken, and stick it on an official-looking card. So scrutinize it carefully, especially if the caller asks to enter. If you are in doubt, take the card, shut the door and phone the company he claims to represent, asking if they have an employee of that name. If he is genuine he will not object but will likely commend you for your prudence.

Do not let anyone into the house unless they can be so identified. If you have to leave the front door to get something, then close it, telling the caller to wait. Again, if he is genuine, he won't mind. Never accompany a caller into the garden or elsewhere and leave the door ajar. If he seems on the level, first get your key, then close and lock the door behind you. You can remark something about children getting in and stealing, if you feel you need an excuse.

Do not be fooled by children: they may look innocent but often are far from it. If strange children call, it is likely to be with an ulterior motive. If one asks for a drink, shut the door, get a drink of water and take it to him, waiting until he has finished. You may well find that he drinks very little because he is not thirsty at all.

Unfortunately, some genuine cases could suffer as a result of the need for such measures. Perhaps an accident victim, or someone that has just been mugged, or even a child who has been molested, may call asking for help. You may thus be put in a quandary: is this genuine or is it just another clever trick?

Obviously if there are injuries and bleeding, the case is genuine and whatever help is needed should be quickly rendered. In other cases, be helpful but cautious. If there seems some evidence to support their story and signs of distress, ask them in, but close the door firmly after them. If you are alone, take them to the room where the phone is installed and phone immediately for the appropriate authority. Thus they will not be out of your sight. Then you can sit down with them and try to calm them until help arrives. By following this procedure, you will help any who need it, yet not expose yourself to unnecessary risk.

It is likely that those who most need the warnings in this section, the frail and elderly living alone, will not be reading this book. If you have elderly relatives or acquaintances, you will be doing them a service to read this to them, and make sure they have grasped the points.

Conclusion

Our examination of the subject of home security ends here for the non-technical reader. We have looked at all the ways a home can be made physically secure, the various items of an alarm system, how they work, which are most suitable for particular cases, how to install and how to find faults. We have taken a look at the gadgetry on offer – which are good and which are dubious. We have also seen how to avoid being taken in by the plausible confidence tricksters. So, the reader should now be able to guard his home against all comers. It is hoped that the book has accomplished this and so played a part in the fight against crime.

Some points to remember are as follows. First, the strength of the chain is in its weakest link, so check your home for its vulnerable points and physically secure *all* of them. Most local police authorities have crime prevention officers who will, if requested, come and survey your premises and advise what security measures are needed. This can be a check on your own survey. Second, protect the perimeter with sensors; use space sensors to back them up. Third, use loud sounders inside and out. Fourth, be wary of gadgets; use principles here outlined to determine their worth. Finally, aim for 'friendly security', so that you will not be tempted to relax it when preoccupied or in a hurry – and never do so.

13 British Standards for intruder alarms

This chapter is mainly for technicians. However, as insurance companies usually require any alarm system to be according to the relevant BS to qualify for discounts, it will enable a check to be made on any proposed system, or one already installed, to see if it qualifies. It should be mentioned that DIY systems are not usually acceptable; they have to be installed by firms with specified trade qualifications.

There are two British Standards that apply to intruder alarms. These set out certain standards which the authority deems to be the minimum. Not all the recommendations have the support of those in the industry, and with most of the controversial ones the standard is considered to be not high enough. These though are in the minority, and the greater number are sound. Many do not apply to domestic systems.

The British Standard that applies to intruder alarm systems is BS 4737; and that which covers wire-free (radio) alarm systems is BS 6799. These are lengthy documents, and numerous modifications and updates have been added over the years. The following is just a summary of the main points, which concentrates on technical details and is up to date at the time of going to press. Anyone having a particular interest in a certain aspect should consult the actual BS, of which most large libraries have a copy.

BS 4737

Part 1

Part 1 deals with the intruder alarm system excluding the sensors.

Housing

The housing of the control unit, outside bells and other equipment should be of mild steel of not less than 1.2 mm gauge, stainless steel not less than 1 mm, or polycarbonate not less than 2 mm. All should have anti-tamper microswitches.

Wiring

When the system is set, an open-circuit or short-circuit of any sensor cable should trigger the alarm. It is also desirable, though not mandatory, that cables should be monitored when the system is not set.

They should be 0.5 mm^2 class 5 cable as described in BS 6360:1981, with PVC insulation not less than 0.25 mm, and sheathed with TM2 compound. Tinsel wire can be used in which case each strand should have a resistance of less than 270 Ω/km at 20 °C with PVC T12 insulation of not less than 0.3 mm. The total resistance of all circuits should be less than that which would reduce the voltage to below the required minimum at full load.

Joints can be made by wrapping, crimping, soldering, clamping, by plug and socket, or by wire-to-wire joint either insulated or in a junction box.

All wiring must be within the protected area or be mechanically protected where this is not possible. None should be run in the same conduit or trunking as mains wiring unless physically separated.

Control equipment

Control units should be within the protected premises and not be visible from the outside. All detection circuits should latch. Each zone should give either audible or visible indication of an alarm condition existing during setting or unsetting, or when testing the system.

It should not be possible to set the system in an alarm condition. Unsetting should be by keys having no less than 200 differs (200 possible key profiles), so the chance of a similar key being available is small. Alternatively another method having a similar security level can be used. (There is no mention at present in the BS of key-pads, although these would be covered by this alternative.) After an alarm, there should be a clear indication of which zone was triggered, and if more than one, which was the first.

Setting and exiting

Exiting should be by a timed delay circuit, or a setting control at or immediately outside the exit point. An audible warning should be heard over the whole exit route and immediately outside it. Shunt locks used for exiting should have at least 200 differs.

Unsetting

This can be by means of a lock-switch or door-switch, or any other means if within the protected area. Audible warning should be given during entry until the system is unset, but this is required only if a remote signalling device is connected to the system. A second delay is permitted to start after the first has expired, during which only a local sounder operates. If still unset after the second delay a full alarm should occur.

If a sensor in any other part of the premises operates during the entry routine, a full alarm should sound.

Circuits can be isolated by a shunt lock having more than 1000 differs, or by any other means of similar security level.

Alarm response

The system should respond to an alarm signal of not less than 200 ms but more than 800 ms. Response should be within 5 seconds. At least one warning device (sounder or remote signaller) should operate within 5 seconds.

If the premises are partly occupied when the system is set, and a sensor is thereafter triggered, an alarm should be signalled by a local sounder not more than 5 minutes later. If the system remains unset, a full alarm should sound not more than 5 seconds after this.

It should not be possible for a system to be reset by a subscriber after an alarm, but only by the alarm company's engineer or an approved trained person.

Power

Power should be derived from the mains via an isolating transformer and correctly fused. It should not be supplied through a plug and socket.

Secondary standby batteries should have a capacity of not less than 8 hours operation, and should be recharged within 24 hours. Changeover in the event of mains failure should be automatic.

Primary batteries should have sufficient capacity to run for not less than 4 hours in the alarm condition and should have the date of installation marked on them. This applies also to secondary batteries that are not automatically recharged. The system operating voltage should be 12 V, but higher voltages are permitted, though not exceeding 50 V.

If the voltage becomes low, an alarm should be triggered. It should not be possible to set the system with low voltage.

Sounders

Housing should be as described for control boxes. In addition there should be no projections to which ropes or chains could be attached.

At least two fundamental frequencies between 300 Hz and 3 kHz should be generated.

Output should be greater than 70 dBA at 3 m. This is measured by mounting the sounder on a solid support with at least 50 mm of the material surrounding it, and a counterweight at the rear. The assembly is suspended 1.2 m from the floor and a measurement is taken on axis at 3 m distance. The reading should be greater than 65 dBA in any direction. Absorbent material should be laid on the floor to prevent reinforcing reflections during the test.

Self-actuating bells should be used having a battery with a capacity of not less than 2 hours sounding. It should be rechargeable within 24 hours. A short-circuit across the charger should not discharge the battery. An alarm should sound if connections are changed or the tamper circuit is actuated.

The bell should be capable of functioning in an environment of –10 °C to 55 °C and up to 95 per cent humidity.

Internal sounders have the same requirements as external ones and should also be self-actuated.

Remote signalling

Any BT line used for remote signalling should be either a dedicated outgoing-calls-only line or ex-directory. It is recommended that it be buried underground or concealed over its whole length. It should be continuously monitored for faults.

For 999 diallers, the equipment should be triggered by any alarm signal longer than 200 ms, and should respond within 5 seconds. Once started, it should not be possible to interrupt the transmission.

If the sounders are delayed, the set delay will be reduced to less than 30 seconds if a telephone line fault is detected by the dialler. An alarm initiated by a panic button may not necessarily operate the sounders at all if it is deemed more prudent to rely on remote signalling to summon help.

Digital communicators should also be triggered by an alarm signal of not less than 200 ms, but the response should be within 1 second.

If no contact is established with the monitoring station within 1 minute, the BT line should be released and the process restarted. If there is no contact after no fewer than 3 attempts, the sounders should operate in less than 10 seconds. If contact is made but no acknowledgement is received after not more than 10 message transmissions, the process should be restarted.

Standby power supply batteries for remote signalling apparatus should be of sufficient capacity to power it for at least 5 alarms if rechargeable, and should be recharged within 24 hours. Primary non-chargeable batteries should be able to power the equipment for at least 20 alarms. The date of installation should be marked on each battery.

Part 2

Part 2 refers to alarm equipment that is used only with deliberately operated sensors (panic buttons). Apart from exit routines, the conditions are virtually identical to Part 1.

Part 3

Part 3 covers the various sensors and their installation.

3/1 Taut wiring

This is used to protect walls and other structures that could be broken through. It should be of hard-drawn copper of 0.3–0.4 mm gauge. PVC insulation should be 0.2–0.3 mm thick.

Fixing points should be no more than 600 mm apart, and the spacing between adjacent runs no more than 100 mm. If the wire is run in tubes or grooved rods, the spacing to adjacent ones should be the same, and they should be supported at not more than 1 m intervals. If recessed into supports, the amount of recess should be more than 5 mm but less than 10 mm from the end of the tube.

The wire emanating from the tube ends should be supported at no more than 50 mm from the tubes. Less than 50 mm of displacement should be necessary to break the wire and sound the alarm.

3/2 Foil on glass

Foil should be less than 0.04 mm thick and less than 12.5 mm wide.

If a single-pole configuration is used, the foil can be laid as a rectangle between 50 mm and 100 mm from the edge of the glass. It can be run as a single strip not less than 300 mm long if the short window dimension is less than 600 mm. In this case it would be run through the centre, parallel to the long edges. It can also be laid as parallel strips not more than 200 mm apart, the ends terminating 50–100 mm from the edge. Unframed glass can be protected with a loop not less than 200 mm by 200 mm.

3/4 Microwave sensors

The area covered by a microwave sensor using the Doppler effect is defined as that in which an alarm is triggered by a person weighing 40–80 kg moving at between 0.3 and 0.6 m/s through a distance of 2 m, or

20 per cent of his radial separation from the sensor. The sensor should respond only to signals of over 200 ms.

3/5 Ultrasonic sensors

The area covered by an ultrasonic sensor using the Doppler effect is that in which an alarm is triggered by a 40–80 kg person moving through a distance of 2 m at any speed. Response should be to signals over 200 ms.

The frequency must be higher than 22 kHz. The sensor should be capable of operating at temperatures of 0–40 °C and at humidities of between 10 and 90 per cent.

A warning is given that nuisance or health hazard is possible from intense ultrasonic radiation. Safe limits are still under consideration, but it is recommended that the device does not radiate when persons are lawfully near.

3/6 Acoustic detectors

The sensitivity control should be set to trigger when a sound is greater than 15 dB over the ambient noise, or at a maximum of 85 dB. There should be no more than 6 dB variation within those limits.

The detector should respond to signals longer than 5 seconds in any 30 second period, or any input greater than 120 dB for 100 ms.

3/7 Passive infrared detectors

The area covered is defined as where an alarm is triggered by a 40–80 kg person moving laterally through a distance of 2 m at any speed. It should not generate an alarm signal when a target that is equivalent to a 40–80 kg person fills the view of the detector, and is heated at an even rate of less than 0.1 °C per second.

When a remote signalling device is included in the system, it should not be connected until there has been at least 7 days trouble-free operation of the PIR, with the exception that the detector is replacing one of similar type or there is an urgent need for immediate maximum security.

3/8 Capacitive detectors

These are sensors that operate when there is a change of capacitance or a rate of change in the proximity of the device. Their area of detection is defined in the same way as for a microwave sensor, namely as that in which an alarm is triggered by a person weighing 40–80 kg moving at between 0.3 and 0.6 m/s through a distance of 2 m.

3/9 Pressure mats

The pressure required to operate a pressure mat is that applied by a disc of 60±5 mm in diameter, pressing at right angles to the mat with a force greater than 100 newtons. It should not be actuated by a force applied by the same disc of less than 20 newtons. (1 newton is the force applied to 1 kg to accelerate it 1 metre per second per second; it is equivalent to 100,000 dynes or 0.225 pounds force.)

3/12 Beam interrupters

An alarm should not be generated if a beam is reduced by less than 50 per cent. The alarm signal should be greater than 800 ms in length for an interruption longer than 40 ms, but should not be triggered for one less than 20 ms. In all cases the source should be modulated to prevent the receiver being affected by another source, accidental or deliberate. (The type and frequency of the modulation are not specified.)

3/14 Deliberately operated sensors (panic buttons)

Three types are described: (1) that requiring a single force on one element; (2) that needing two simultaneous forces on two different elements; and (3) those operated by two consecutive forces applied to different elements, the first being maintained while the second is applied. The sensors can be latching or non-latching.

For a manually operated device, the force required should be within 4–5 newtons as applied by a 6 mm diameter disc. For pedal-operated devices the force should be within 5–8 newtons applied by a 12 mm diameter disc.

3/30 Cables

Cables for intruder systems should be either plain or tinned annealed solid copper not less than 0.2 mm^2, with a maximum resistance of 95 Ω/km. A pro rata resistance should apply for cables up to 0.5 mm^2.

Stranded cables should be not less than 0.22 mm^2. Any strand joints should be brazed or hard soldered and not be less than 300 mm apart. The tensile strength of these should be not less than 90 per cent of adjacent continuous cable.

Cores should be no fewer than 7 strands of 0.2 mm conductor. Insulation should overlay the wire but not adhere to it and so prevent a clean strip. Sheaths should likewise not adhere to cores. Cable sheaths should be not less than 0.4 mm thick. Insulation resistance should be greater than 50 MΩ for 1 kM at 20 °C.

Part 4

4/2

This deals with the security of the installing company, and maintenance. Staff must be thoroughly vetted before being taken on to deal with clients' security systems, and must receive adequate training to ensure competence. The confidentiality of all records and details must be totally preserved.

Maintenance visits and tests should be made not more than every 12 months for mains-operated systems using local sounders. For battery-powered systems, the maximum time between visits should be 6 months. When remote signalling equipment is included, the visits should also be not more than 6 months apart.

On each visit the following should be carried out: check of record book, for problems arising since the last visit; visual check of the system and any changes in layout or stock storage that could affect it; test of all sensors; examination of flexible cables; check of power supplies including standby batteries; check of control equipment; check of sounder; check of remote communicators; and a full operational test.

Accurate records of all alarm events, maintenance and repairs should be kept.

BS 6799

This standard describes wire-free (radio) alarm systems. There is no mention of the frequencies or the radiated powers to be used, as these are regulated by the Department of Trade and Industry. At present the DTI has allocated 173.225 MHz for use by intruder alarms providing the transmitter is crystal controlled at 25 kHz channel spacing. No licence is required for an effective radiated power (ERP) of up to 1 mW. For longer ranges or in electrically noisy environments, up to 10 mW is permitted, but this requires a licence. Another suitable band for longer range is 458.5–458.8 MHz, in which up to 500 mW ERP is permitted without a licence. These bands are for transmission of data and not for speech.

BS 6799 describes five levels of security in systems using wire-free sensors. The need for such a choice arises from the power supply limitations. Sensors need to be reasonably small, yet each must contain a battery having a life of some 6 months. Continuous transmission to provide 24 hour monitoring is thus not feasible. The five different levels reflect different compromises between monitoring and battery life:

1 The sensors transmit only when actually triggered and also signal when the battery is getting low, having capacity for only 7 further days of operation.
2 As 1, but the sensors also transmit a code identifying which one has been triggered. This indicates which part of the premises is affected.
3 As 2, but the receiver monitors the channel for interfering or blocking signals lasting for more than 30 seconds, and gives a warning of such.
4 As 3, but the sensors transmit a return-to-normal signal after an alarm. They also regularly report by transmitting their status and battery condition at no longer than 8.4 hour intervals. If no status report is received in an 8.4 hour period the receiver signals a fault condition; if there are no reports for three consecutive periods a full alarm is sounded. A fault indication is given if a low-battery signal is received in two consecutive reports.
5 As 4, but the reports are transmitted at 1.2 hour intervals.

The standard also outlines the advantages and disadvantages of radio sensors. The advantages are the obvious ones of eliminating wiring, no damage to historic decor, portability, and easy addition or changes to the system. The disadvantages are the possibility of blocking or interference, limited monitoring and no anti-tamper protection, as well as frequent battery replacements for each sensor.

Much of this does not apply to the normal domestic system: for example, the anti-tamper circuits described are not necessary, as we have seen. In some cases a somewhat lower standard of security may be deemed sufficient for the circumstances in order to obtain 'friendly security' and minimum inconvenience to the occupiers. So it may be considered an advantage not to have a sensor on the front door if it is robust and well protected with deadlocks, and so avoid the hassle of exit and entry procedures.

14 Sureguard alarm systems

This chapter is intended for those with some knowledge and experience of constructing simple electronic circuits. The two Sureguard circuits here described are quite easy to build, yet offer features not found on expensive commercial systems. On the other hand they are not cluttered with features that will not be needed in the applications they were designed for.

Of the two models described, the first is the Sureguard Mark 8 which is intended for home installation. The second, the Sureguard Mark 9, is designed for public halls, although it could also be used as an alternative to the Mark 8 for the home. The basic circuit is the same for both, but they have different features.

Sureguard Mark 8

As with most other alarms, there is an open-circuit facility to which pressure mats and other normally-open sensors can be connected. There is also the usual closed loop for the series connection of magnetic switches, PIR devices and other normally-closed sensors. This has about a half-second delay so that momentary actuation of sensors by heavy vibration or other causes will not create a false alarm.

A feature of the closed loop is its very low standing current. Many systems circulate loop currents of several milliamperes; with some of the older ones it was tens of milliamperes. This imposes an appreciable drain on standby batteries if the mains supply fails. As a result, large and expensive batteries such as the HP1 are required.

The Sureguard loop in both models circulates just 0.1 mA. This enables less expensive batteries to be used, which are also more readily available; two lantern batteries connected in series to give 12 V is recommended. These have been found to last up to two years with normal domestic use without using the mains supply at all.

A mains supply circuit has been included, however, to power any PIR devices that may be fitted. Without these, the mains supply circuit could be dispensed with, and the system run solely on batteries. This is thus an option for a system to be installed where there is no mains supply. If though there were many lengthy bell soundings due to false alarms, the battery life would be considerably reduced. When the alarm is operating from the mains, the switch-over to batteries is automatic if the mains fails.

Testing

It is in the test facilities that the Sureguard Mark 8 is possibly unique. Many systems test only the continuity of the loop, having an indicator lamp that lights up if the loop is complete. Some test the internal control and latching circuit by switching a lamp in place of the bell, then triggering the circuit.

All these leave the most vulnerable part of the system, the bell, untested. Actual sounding tests should be made with all systems, at periodic intervals, say once a month, but in practice rarely are. Even if they are made, something could happen to the bell or its wiring between tests. This would leave the system useless, and the home vulnerable for the period until the next test.

The Sureguard Mark 8 has a comprehensive check which tests everything including both bells without actually sounding them, and this is done with just one action on the part of the user at each switch-on. The system is thus fully tested at every switch-on – even late at night, and without disturbing the neighbours.

As two bells are tested, two must be used; if only one is connected to the system, a fault will be indicated when a test is made. As shown in previous chapters it is highly desirable to use two bells, the second mounted inside the premises, or at the back of the house. A third bell can be connected in parallel with the second, but this would not be independently tested.

Panic buttons

Another feature is the panic button system. With many control units the system does not latch when a panic button is operated. So the panic button must latch mechanically to keep the alarm sounding. Afterwards it must be reset by further depression and release, or by means of a key. With Sureguard the system itself latches so the button can be a simple non-latching door-bell press. A slender one can be chosen to fit inconspicuously to a door frame, under a desk, or any convenient position. Any number can be fitted in parallel, and all will actuate the alarm whether it is switched on or not. Cutting the button wire will not stop the alarm once it has started.

Exit circuit

This is connected to the exit door and controls only that door. So the alarm can be switched on for any length of time before leaving, and unexpected delays while going to the exit door cause no problems. Once the exit door is opened, an alarm will be triggered in about 8 seconds if the door is not shut. This allows ample time to get through and shut it. If something prevents one leaving after opening the door, such as the phone ringing, shutting it cancels the operation. On reopening, the timer starts again and the full exit time is available.

The same applies when re-entering. If the door is shut immediately after entering, there is no hurry to switch off. So if, on entering, the phone starts ringing, it can be answered and the alarm switched off later. Of course all other sensors will still be on guard, so care must be taken not to actuate any of them while the system is still on.

If an intruder breaks in through the exit door he is unlikely to shut it within 8 seconds, but even if he does, he will undoubtedly trigger another sensor such as a pressure mat or a PIR device. As we have seen, break-ins are rarely through front doors anyway, if they are physically robust as they should be. So, with very little compromise in security, the exit arrangement is one of the most user friendly for a wired system, and was designed with this objective.

Control switch

In this the system departs from current practice. A key-operated switch is not used, just a three-way rotary switch with a knob. There are several reasons for this. One is the requirement for friendly security. Carrying yet another key, and fumbling for it every time you switch on or off, is a chore, and it leads to the temptation not to bother when popping out for a short while. Yet this is the very time when many burglaries take place. So a key-operated system can actually compromise security.

The best security for the control unit is to conceal it, as shown in earlier chapters. One that is prominently on view can invite an attack which it may not survive, key-operated or not.

If any qualms are felt about the security of a simple knob control, a dummy key-switch could be fitted to the front panel, and the actual control mounted unlabelled on the side or even underneath, out of sight.

Another reason for not using a key-switch is a practical one. The comprehensive Sureguard test facility requires a four-pole switch, and key-switches are not generally available with this number of poles.

Sounders

The relay will switch up to 2 A resistive, but of inductive loads such as bells, this must be down-rated. It will though be in excess of 1 A, the rating for most systems, so any sounders, bell or siren, can be used up to that current rating. It must be remembered though that high-current devices would soon exhaust standby batteries if they were run from them for any length of time.

The recommended bell is the Tann Synchronome B6D12 (150 mm) or the B8D12 (200 mm). These can be mounted in the open without a box, have a very loud and strident tone and also have a low current consumption of less than 0.1 A.

Radio interference

Both Sureguard models have radio suppression which greatly reduces the possibility of false alarms due to a nearby strong radio transmission. This also prevents self-oscillation of the control transistor which has been a problem with some alarm systems.

Reliability

This was a special consideration in the design of the Sureguard models. All electronic components are liable to failure, and a failure rate per thousand samples, plus mean time to expected failure, are actually specified by component manufacturers and are the subject of a British Standard as well as an IEC standard. The active components such as transistors are more vulnerable than passive ones such as resistors and capacitors.

It follows that the more components a unit has, the more likely statistically it is to break down, especially if they are active components. Many commercial units now use dedicated chips, that is integrated circuits specifically designed for the purpose. These contain a large number of transistor elements and, although few actually do fail, the possibility is always present. For security purposes, a zero failure rate, though statistically unobtainable, is the goal to aim for.

The Sureguard units were designed with this aim, so all their features are obtained with an absolute minimum of components, especially active ones. In fact only one transistor is involved when an alarm is triggered, and even that is immediately bypassed as soon as it has responded. Latching is accomplished by a totally enclosed relay which is a very reliable component when operated within its design specification.

Other components, resistors and diodes have high reliability factors but, even so, these have also been reduced to the minimum. Electrolytic capacitors are more prone to faults, but those used are in the timing circuits, and any fault in them will not affect the ability of the unit to trigger an alarm.

Bell timing

It is now considered desirable, and is mandatory in some areas, that an alarm should sound for no more than 20 minutes. Certainly intruders would not be still on the scene after that time, and sufficient warning will

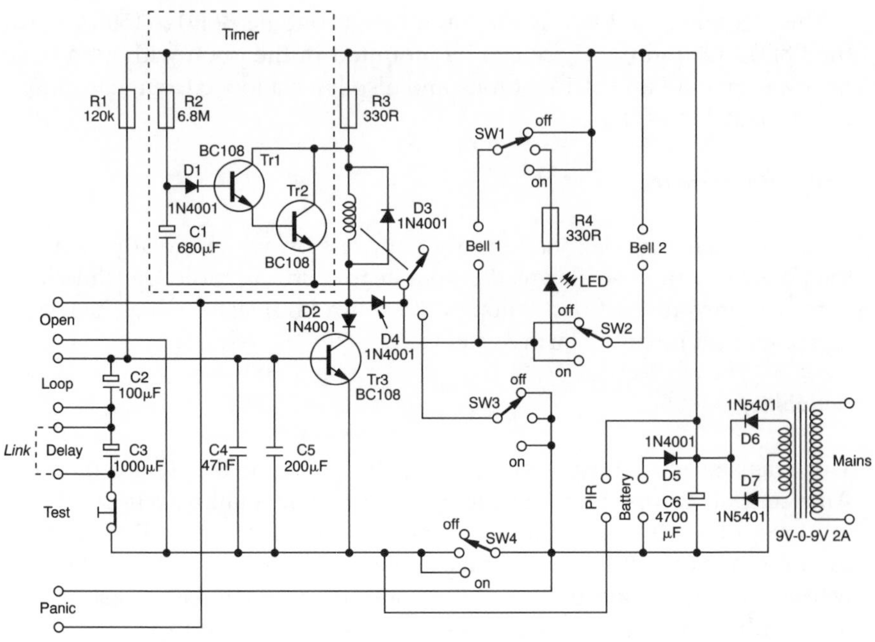

Figure 62 *Sureguard Mark 8 circuit diagram.*

have been given that a possible intrusion has occurred. So modern alarm systems have a timer that switches off the sounders, or switches off the complete system and resets it.

The design of such a timer is by no means difficult, but it usually entails an active electronic switching element placed in series with the bell or supply. It is evident from what has just been said that this must increase the risk of failure. If this element or any associated controlling component fails, it will put the entire system out of action.

The timer devised for both Sureguard models has no series elements at all. It shunts the relay coil and is not effectively in circuit until the time has elapsed. Then it forms a low-resistance shunt path across the coil, so delatching the relay and resetting the system.

Technical description

On consulting the Sureguard Mark 8 circuit diagram (Figure 62), it can be seen that forward bias for the base of the control transistor is supplied via R1, but the base is grounded through the loop and so the transistor is non-conductive. If the loop is broken, voltage appears on the base through R1, and the transistor conducts, so energizing the relay.

When thus energized, the relay contacts close and connect the bottom end of the relay coil to the negative line, thereby latching it on. It also completes the negative circuit to the bells, causing them to sound.

The capacitor C2 across the loop terminals provides the 0.5 second delay, and C3 across the exit terminals gives the 8 second delay. Delays can be varied by altering these capacitances: an increase in value increases the time delay. If the exit delay is not used, a link must be connected across the delay terminals. Note that the series test button is a press-to-break type.

If a sensor such as a pressure mat short-circuits the 'open' terminals, the bottom end of the relay coil is thereby connected to the negative line, so energizing the relay, latching it on, and sounding the alarm. The transistor is not in use in this case, so in the rare event of the transistor failing, pressure mats would still be operative.

The panic button works in the same way as the pressure mat, only the negative return is taken directly to the supply negative, so bypassing the switch. It thus works irrespective of whether the switch is on or off.

The main control switch is a four-pole three-way type. When it is in the 'test' position, SW1 and SW2 connect the two bells and the LED in series. A small current, sufficient to light the LED but insufficient to sound the bells, passes when the circuit is triggered, indicating that the bell circuits are continuous. If there is a break in either, the LED does not light. In the 'off' and 'on' positions, the bells are connected in parallel ready to sound. This is necessary in the 'off' position so that they will sound if the panic button is used.

It may be wondered just what the function of SW3 is, which has all its contacts joined together. The switch is of the break-before-make type, so when the switch is turned from one position to another, the relay, if latched, is allowed to delatch. Otherwise when turning from 'test' to 'on' the alarm would start sounding, and it would also be impossible to switch off!

SW4 switches the negative supply so that it is closed for testing and for the system 'on' position. All switching apart from the series/parallel bell circuits is in the negative side of the supply.

The mains-to-battery changeover is of interest as it involves just one diode D5, so maintaining the minimum possible component count and maximum reliability. When the mains supply is present, the voltage on the diode cathode is around 13.5 V which, being higher than the 12 V battery voltage on the anode, makes the diode non-conductive. The battery is thus isolated and does not discharge current into the circuit. If the supply fails, the diode starts conducting as there is then no positive voltage on its cathode, so the battery supplies current to the circuit. With resumption of the mains supply, the conditions revert to their former state and the diode again isolates the battery.

The mains power unit is a conventional full-wave circuit. A 9–0–9 V transformer secondary gives 12 V DC on full load, but around 13.5 V with the low current taken by the Sureguard circuit.

Timer

As mentioned earlier, this is a shunt actuating circuit to avoid the reliability hazard introduced by a series device. The circuit is shown within the dotted box on the main circuit diagram. When the relay contacts close, voltage is applied to the timer circuit and C1 starts to charge slowly through R2. The voltage on the base of the first transistor also slowly rises until it starts to conduct and supply base current to the second transistor. This passes amplified current which causes a voltage drop over R3 in series with the relay coil. The voltage across the coil thus drops until it is no longer sufficient to hold the contacts, and the relay delatches.

The capacitor discharges back through R2 and the bell circuit, and also the base/emitter circuits of the transistors. Discharge is thus quicker than the original charge. If the alarm is triggered again before the capacitor is fully discharged, the alarm will sound but for a shorter period, depending on the state of discharge. The original timing depends on the type of relay used and its release voltage rating, the current gain h_{FE} of the transistors, and the precise value of the electrolytic capacitor which has a wide tolerance. However, the timing is not critical and should be between 10 and 20 minutes. Anything much outside these limits can be corrected by changing the value of the capacitor.

When checking the timing, it is not necessary or desirable to have the bells sounding! This is avoided by switching the main control to the middle 'test' position; then trigger the alarm and the LED will light until the time expires. Bells will need to be connected for this test.

Operating Sureguard Mark 8

Operating is very simple, this being one of the required design features. First, turn the control to the centre 'test' position. If the LED lights then the loop is open-circuit, which means a door or window is not properly closed, or something is standing on a pressure mat. Turn back to 'off', and find and correct the matter.

If the LED remains off, press the 'test' press button. The LED should now come on, this indicating that the loop and control unit are properly functioning and both bell circuits are continuous. Now turn slowly to 'on', and that's it: the system is fully tested and armed. If you turn too quickly, the relay may not have time to delatch from the test, and the alarm may sound. If it does, turn quickly back to 'test' and try again – more slowly. In most cases, though, the relay will beat you to it.

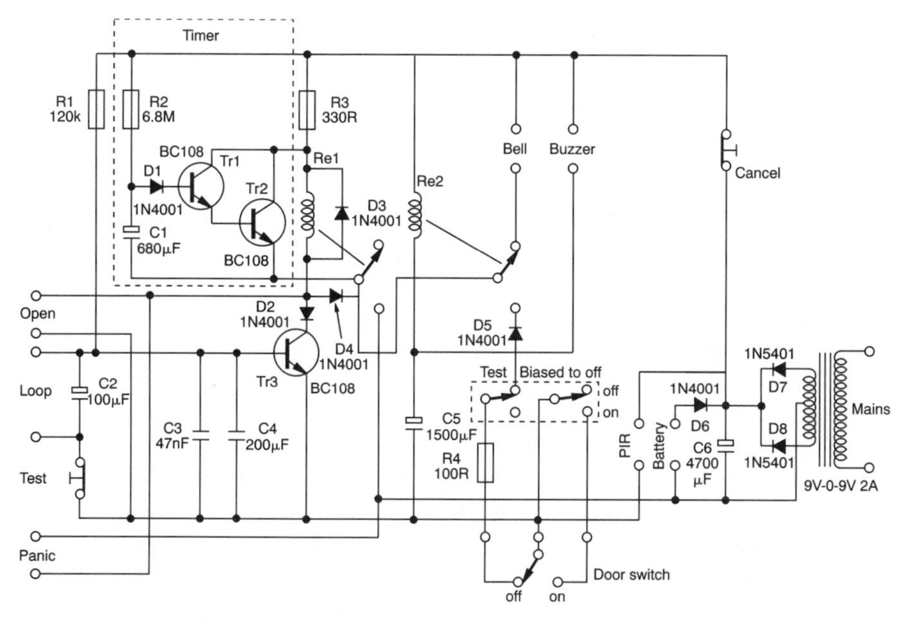

Figure 63 *Sureguard Mark 9 circuit diagram.*

Sureguard Mark 9

The problem with public halls and similar locations is that with so many different people coming and going, the alarm system can be, and often is, misused or ignored and not switched on for much of the time. Sureguard Mark 9 was designed to overcome this problem; the whole system is switched on and off by just locking and unlocking the exit door. Thus it is always on when the door is locked and there are no difficulties with an exit procedure.

A buzzer is mounted near the exit door so that it can be heard from the outside. When the door is locked by means of a lock-switch, the buzzer sounds and dies away. If a door or window fitted with a sensor has been left open, the buzzer sounds continuously.

The control unit uses a similar circuit to that of the Sureguard Mark 8, so what has been said about that also applies here. It is shown in Figure 63. There are two differences: one is that there is no exit circuit as this is not required; the other is that the comprehensive test feature is omitted. As the switch-on is automatic, there can be no manual testing. While this reduces security to a degree, it is better than having the system not switched on at all for much of the time, as often happens with conventional units in this situation. However, periodic testing is provided for.

Two, three or more bells can be fitted, and these are all connected in parallel, unlike the Mark 8 bells which have to be connected to separate circuits to enable the bell test to be made.

Technical details

A single-pole double-throw lock-switch must be installed in the exit door either in place of, or in addition to, the existing lock. An extra relay, Re2, is used which switches from bell to buzzer and back.

When the system is switched on, current flows through C5 which is in series with the buzzer, which sounds for a few seconds until the capacitor is charged. The relay coil which is in parallel with the buzzer is also energized, but is released when the capacitor is charged. So if a sensor is then actuated, the bell will sound.

If a window or door is open, the first relay energizes and is latched on in the normal way. This bypasses the capacitor so that the second relay stays on and so the buzzer sounds continuously.

If all is well, the second relay returns to rest after switch-on, in which position the bell is switched in circuit in place of the buzzer. So if a sensor is then actuated, the bell will sound. After an alarm or a continuous buzz, the system can only be switched off by pressing the 'cancel' button which interrupts the supply and so delatches the relay. In the 'off' position, the door contacts discharge C5 through R4, a current-limiting resistor, so that it will be fully discharged and ready for the next switch-on.

For testing, the test button open-circuits the loop as with the Mark 8 version, so triggering the detection circuit. The double-pole double-throw test switch is biased to 'off' to prevent someone leaving it in the wrong position, which otherwise is likely. This bypasses the door switch, switching the system on, and open-circuiting the capacitor discharge shunt.

Operation

As previously described, this is just a case of locking the exit door. If the buzzer sounds continuously, unlock and re-enter, cancel the alarm by pressing the cancel button, then find which window or door has been left open. When the buzzer sounds and dies after locking, the system is in order and on guard.

Because the Mark 9 lacks the test facility of the Mark 8, a test should be carried out regularly, say once a month, by someone made responsible for it. This will normally be the caretaker. The procedure is as follows:

1 Close all protected doors and windows. Push the biased switch on. The buzzer should sound and die away. Release the switch.

2 Press the test button, hold it down and push the biased switch. The buzzer should sound continuously. Release the button and switch. Press the cancel switch to stop the buzzer.
3 Push the biased switch, hold it on and then press the test button. The bells should sound. Release the button and switch. Check that all bells are in fact working. Press the cancel button to stop.

Construction of the Sureguard circuits

With so few components for the Mark 8, most of the wiring is to switches and terminals. It is hardly worth making a printed circuit; in fact they can introduce an element of unreliability, especially the matrix type. Hairline cracks that produce intermittent breaks are quite common, as are slivers of solder bridging adjacent tracks.

It is worth noting here that printed circuits were originally introduced by manufacturers to save money. Once designed and produced, they saved a considerable number of working hours otherwise spent in hand wiring a large number of identical circuits, especially when these were complex such as TV receivers. When it comes to one-offs, though, especially for a relatively simple circuit, hand wiring is usually quicker and more practical. Yet many electronic hobbyists seem to think that if a printed circuit is not used, the job has not been constructed properly. In cases where DIL integrated circuits (ICs) are used, a printed circuit is the only practical way of mounting them.

With the Sureguard Mark 8, apart from the existing terminals, a few extra tags mounted at convenient positions in the control box should support all the components except the relay. This can be glued with epoxy resin to the box by its top so that the terminals are uppermost.

The Mark 9 can be constructed using a twin 18-tag tag-board. The layout is shown in Figure 64. The relays are shown larger in proportion to the board than they actually are, in order to make their connections clear. They can be glued to the board with tags uppermost. Other components are shown slightly smaller so that the wiring can be clearly seen.

Components for Mark 8

R1, 120 kΩ; R2, 6.8 MΩ; R3, 330 Ω; R4, 330 Ω.
C1, 680 μF; C2, 100 μF; C3, 1000 μF; C4, 47 kpF; C5, 200 pF non-inductive; C6, 4700 μF, at least 25 V DC, 2 A ripple current.
D1–5, 1N4001; D6–7, 1N5401.
LED indicator.
Tr1–3, BC 108.

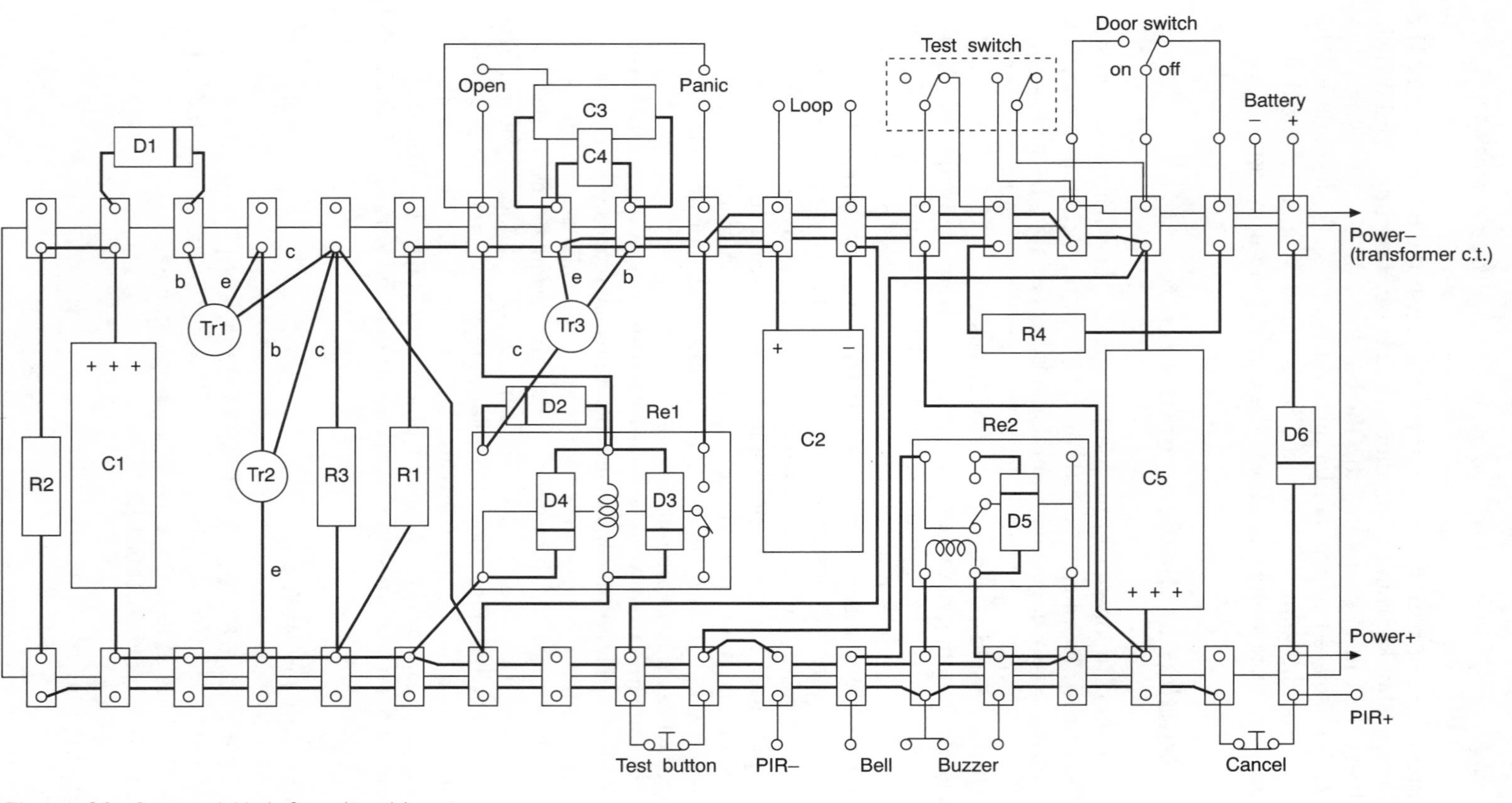

Figure 64 *Sureguard Mark 9 tag-board layout.*

Re1, Radiospares 346–665.
Control switch, three-way four-pole, break-before-make.
Test button, single-pole, press-to-break.
Mains transformer, 9–0–9 V 2 A secondary.

Components for Mark 9

R1, 120 kΩ; R2, 6.8 MΩ; R3, 330 Ω; R4, 100 Ω.
C1, 680 μF; C2, 100 μF; C3, 47 kpF; C4, 200 pF non-inductive; C5, 1500 μF; C6, 4700 μF, at least 25 V DC, 2 A ripple current.
D1–6, IN4001; D7–8, 1N5401.
Tr1–3, BC 108.
Re1, Radiospares 346–665; Re2, Radiospares 348–510.
Test switch, double-pole double-throw, biased.
Test and cancel buttons, single-pole, press-to-break.
Twin 18-tag tag-board.
Mains transformer, 9–0–9 V 2 A secondary.

Remarks on components

Most of these are readily available. The control switch for the Mark 8 model must be a break-before-make, otherwise the alarm cannot be switched off.

Relay 1 (Re1) for both models has rhodium-plated contacts in a sealed gettered chamber, and has a life rating of over 100 million operations. It is rated at 5 V, but has an operating range of 3.75–16 V. The reason for choosing this type, apart from its high reliability, is the low voltage at which the relay is released. This gives a timing period of the required length. Most relays release at a higher voltage and this would considerably shorten the sounding time before the alarm cuts off. If another type of relay is used, it must have a similarly low release voltage, and be sealed for reliability.

The important factor for relay 2 (Re2) in the Mark 9 version is its coil resistance of 320 ohms; any substitute should have a resistance of this order. Being shunted across the buzzer, a low-resistance coil will reduce the sound volume and the sounding time.

The DPDT test switch for the Mark 9 should be biased to prevent it being left in the test position. If not available locally, one can be obtained (type TM22B1) from Cricklewood Electronics, 40 Cricklewood Broadway, London NW2 2YP.

15 Light pollution

We have all encountered pollution, in the air we breathe, the food we eat, the water we drink, and the sea we bathe in. There is the neighbour's hi-fi, and the noise from motor bikes, pneumatic drills, and low-flying aircraft; but what is *light* pollution?

Light pollution is unnecessary excessive light that causes annoyance to others. It can take one of three forms: *sky glow*, seen over major cities; *glare*, the uncomfortable brightness of a light source when seen against a dark background; and *light trespass*, the encroaching of light beyond the boundary of the property containing the source. It can deprive someone of sleep when a bedroom is illuminated by outside light, and it can and does rob us of one of the most awesome and beautiful sights in all nature, the starry night sky.

The first of these is obvious. While some people can sleep undisturbed in bright illumination, many find that it does affect their rest. This is especially so if the light is intermittent, such as a flashing beacon or advertising sign. The source does not have to be nearby, a powerful light in an otherwise dark area can strongly illuminate rooms at some distance and be a cause of annoyance.

The second is now scarcely realised, because very few who have been brought up in an urban area have ever seen the glories of a night sky free from light pollution. It is sad that most people just do not know what they are missing. However, those old enough to remember the wartime blackout, or anyone who has enjoyed a camping holiday in the wilds away from civilisation, know full well what a night sky can look like, and mourn its loss in the baleful, orange streetlamp glare that surrounds our cities at night.

Astronomers in particular bewail the effect. Astronomy is now not just a small minority interest, a growing number are taking it up and acquiring telescopes and other equipment to explore the night skies. But observation of distant galaxies and other faint objects is bedevilled by the ubiquitous orange glare. Often those living in towns and cities have to transport themselves and their bulky equipment to some country site where glare is at a minimum every time they wish to observe.

Little can be done by the ordinary person to reduce streetlight pollution other than lobbying of local councillors. Some enlightened councils are aware of the problem and are implementing the recommendations of the Institution of Lighting Engineers. Among other things, these include

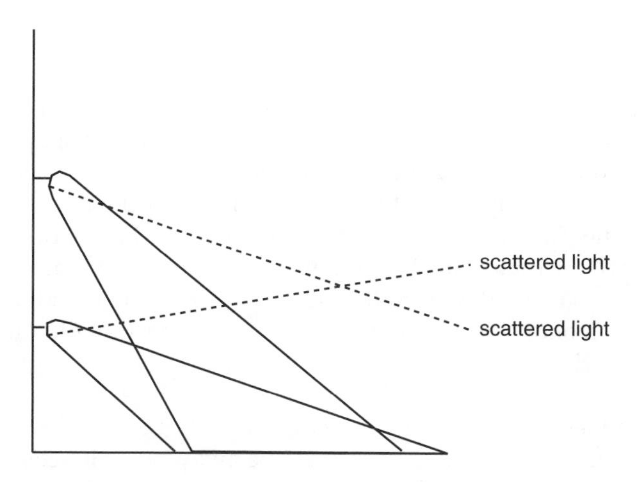

Figure 65 *A high-mounted security light will cover the same area as a low-mounted one, but the scattered light is directed downwards. No light should be propagated above 70° from the perpendicular.*

replacing existing lamps that spill light in all directions with newer types that direct all the light downwards. This means that more light is available for the road where it is needed, and so less powerful lamps can be used. This in turn saves massive street lighting electricity costs and by consuming less, reduces pollution caused by electricity generation. Other councils are niggardly in their allocation of funds for this work and so continue to pay high electricity costs, while continuing to generate light pollution, so robbing all, and especially the younger generation, of a precious resource – a starry, dark night sky.

There is now a Campaign for Dark Skies (CfDS) that is active in promoting the cause by bringing light pollution and its remedies to the attention of the authorities. Lobbying is taking place at all levels including the European Parliament, so it may be in the not too distant future that a directive will arrive compelling local authorities and others to take the necessary measures to eliminate light pollution. Such authorities would be wise to get their act together beforehand!

While most of the light pollution is caused by street lighting, much can also be caused by security lights. Car park lights at halls, pubs, community centres and the like, can add significantly to the pollution in their respective areas. Domestic security lighting can also be a cause of annoyance and frequently is, although the householder may be unaware of it. What then can be done about it?

Timing

The continuous operation of external security lights is in most cases unnecessary and wasteful. As pointed out in an earlier chapter, a light that comes on when an intruder approaches is far more effective, and it can be easily and cheaply achieved by using the PIR detector-operated light. For one thing it warns the householder, if at home, that there is a prowler. A light that is on continuously gives no such warning.

The time that a PIR-controlled light stays on can be adjusted from around 15 seconds to several minutes. It may be thought that a long period 'on' will give greater security, but this is not so. A light adjusted to give a short illumination period will come on again immediately the intruder moves within the PIR detector range. If it goes off and stays off, the indication is that the intruder has gone away, but if it keeps going off and on, he is still around. So more information as to what is happening outside is given than with a PIR that is set to stay on for many minutes.

The rule is then to use a PIR for all outside domestic security lighting, and to set it to the minimum time setting. Apart from being more effective, any light spill into adjacent bedrooms will be of short duration and will cause the minimum of annoyance to neighbours.

One of the few situations where continuous lighting is required is for a car park where there is frequent coming and going and a light going on and off could distract drivers. Decorative floodlighting and advertising signs are usually continuous, but the Institution of Lighting Engineers recommend a switch off no later than 11:00 p.m.

Power

Another misconception is that the brightest light possible should be used for maximum effect. If a large area is to be covered such as a car park, this may be so, but even here there is a downside to having too bright a light. The brighter the light, the deeper the shadows for intruders to lurk. The shadows are not actually darker, but the bright light causes the pupils of an observer to contract, and thereby make the shadows seem darker. The light should be sufficient to illuminate the scene well, and also to enable objects in any darker areas to be easily distinguished.

Too great a power can cause light pollution even if the lamps are correctly angled, as described in the next section. While direct light from the lamp may be controlled, the illuminated object itself will radiate light in all directions. If you can see to read in the reflected light, it is too bright.

For domestic PIRs 150-watt lamps should be more than adequate to illumate porch areas, garage frontages and similar areas, but the CfDS recommend no more than 40 watts for porch lights.

Angles

A major factor in the prevention of light pollution is the angle at which the security light is set. The aim should be that no light is allowed to escape upwards or indeed above the horizontal. The maximum angle from the perpendicular recommended by the Institution of Lighting Engineers is 70°. Generally, the higher the lamp is mounted, the smaller the angle to cover the same floor area. In most domestic installations a point between the ground and first floor is ideal. This also will facilitate the wiring which can be laid under the floorboards of the first floor.

The range should not go beyond the front boundary of the premises to the adjacent road, as it could dazzle and startle drivers if it came on unexpectedly as they are approaching.

While the range of the lighting should be restricted, that of the PIR detector can be more extensive if required, though not so great as to be triggered by passers-by. Most combined PIR security lamps have the PIR detector independently adjustable. Thus, the lamp can be aimed to illuminate the immediate vicinity of the premises, but the detector can range a little further, to perhaps be triggered by someone entering the front gate.

The illuminating of advertising signs should always be done from above, with the light aimed downwards. With the floodlighting of buildings this may not be possible, but the lighting should be carefully aimed so that it just reaches the top of the building but no further. Local sports grounds that are illuminated should have the lighting mounted high and covering only the ground area.

These measures should go some way to eliminating light pollution to the benefit of all, but with no compromise on security.

16 Guard against fire!

Your home is much more likely to be burgled than to suffer a serious fire. However, the consequences of fire can be far more devastating. Major damage to the home and its contents is inevitable, and loss of life is all too likely. So protective measures are even more important than against intruders. While an alarm system is very desirable, it is far better to prevent a fire from starting in the first place.

Causes of fire

Smoking is a very common cause of fire, especially with elderly people. They sometimes drop off to sleep while smoking, and the lighted cigarette drops into the armchair or on to the carpet. It is perhaps futile to urge them to give up, as the habit of a life-time is not easy to break, and they are unlikely to be impressed by arguments involving health, having lived for so long without apparent serious ill effect. The only precaution here possible is to ensure that all furnishings are fire-proof. All upholstery of recent years has had to be of a fire retardant quality, but the armchair in question may be an ancient favourite – and vulnerable. So strong persuasion by relatives may be necessary to effect a change to a more recent model. This may be less than easy, as old armchairs are always more comfortable than any modern one – or so they say!

Younger smokers may not be in danger from this hazard, but the disposal of butts is likely to be the danger. Wastepaper baskets are definitely not the proper receptacle. A butt that appears dead may be hot enough to ignite any inflammable material nearby, especially paper. Many fires have been started by this cause. Ash trays of some sort should always be provided in every room where smoking is likely, and they should be emptied regularly; an overflowing ashtray is itself a hazard.

Kitchen mishaps are another cause of fire. Fat or oil boiling over from the chip pan onto the hob is a common one. Should this happen, do not attempt to carry the blazing pan ahead of you outside, as the draught of the forward motion will blow the flames toward you, or they could ignite other things on the way. If an outside door is close, the pan could be carried outside by backing out, carrying the pan behind, the flames will then stream away from you. The best remedy is a fire blanket thrown over the pan to smother the flames. A foam or powder extinguisher is also effective, but water should not be used as the flaming oil will float over it and so spread.

Another, though less common, cause is for food to boil dry and then burn in an unattended saucepan. Usually though, the burning does not generate flame and is confined to the pan. Water can be used if oil or fat is not involved.

Electrical appliances, such as electric blankets are one common cause of fire. They should be periodically checked by the maker according to his instructions, and if any part of the blanket becomes worn or damaged, it should not be used until professionally checked and repaired. An electric blanket should never be left switched on when folded up, nor should objects such as clothing be placed on the bed. Overheating of the area beneath the object will result.

The TV set is another common cause of fire. Some internal parts get very hot. The makers ensure that they are well ventilated, so it is important to ensure that the ventilation slots at the back of the set are not obstructed. Never site a TV set close to source of heat such as a wall radiator. Unfortunately, faults can occur which lead to parts overheating, and a fire can develop after the set is switched off and the family have gone to bed. As such faults may give no visible sign of trouble, it is a good practice to feel the back of the set before going to bed every night to ensure that the set is quite cool, or that it is cooling if recently used.

Toasters are also responsible for minor fires that can quickly spread. If the toaster fails to switch off and eject the bread, it can catch on fire, and ignite surrounding objects. It is wise not to stray too far away for too long while toast is being made. Should the worst happen, the first thing to do is to switch off at the mains socket, and then quickly smother the toaster with a damp kitchen towel.

Overloaded mains sockets are another possible cause of fire. There are some misconceptions on this subject though. Using a couple or more multiplugs to run a number of appliances from a single socket is not necessarily overloading it. Although not the ideal, it is not dangerous unless the total current of all appliances exceeds 13 amps, which is 3 kilowatts of power. A TV set, video recorder, computer, table lamp and hi-fi, are well within that limit and so could be run with multiplugs from a single outlet socket without danger. In any case they would not all be on at the same time.

Overloading would occur if the the total wattage exceeded 3 kilowatts; a two-bar electric fire or radiator plus one or two other items is approaching the limit; a TV or table lamp should be the only other appliances that could be safely connected in this case.

A danger could arise though, if one of the multiplugs became partly pulled out by the weight of the others, so exposing bare pins. Any metal contacting these could cause a short-circuit and a fire. Ensure all plugs are pushed well home and are firm in their sockets.

It is fact that 13-amp plugs themselves can pose a fire hazard. It is not uncommon to find that a plug is warm after a period of use. Sometimes if the current taken by the appliance is high it can get quite hot; this should not be, the plug should remain cold. It is the live pin that usually gets hot, often too hot to touch. The heat is generated by a loose connection causing arcing inside the plug. It can be due to a loose fuse, in which case, the metal grips that hold the fuse should be squeezed up so that they grip the fuse tightly. When this is done check the plug again after use, if it still gets warm, the fault is due to a loose rivet holding the fuse contact to the pin. If you are handy with a soldering iron, solder the contact to the pin around the rivet to make a sound connection, but if not, discard the plug.

Another important point is to always use the correct fuse for the appliance. If a fuse blows *never* replace it with a nail, pin or other metal object. The fuse is not just a nuisance put there to annoy, it is a safety device to cut off the current if the connected appliance is faulty and taking excess current.

To avoid the possibility of such dangerous improvisations every house should have a stock of several fuses of the common ratings, and keep them where they can readily be found. Most small appliances up to 750 watts should have a 3-amp fuse; up to 1000 watts, 5-amp; 2000 watts, 10-amp; and 3000 watts, 13-amp. It follows that the 3-amp rating will be the most used.

When an appliance is dead and the fuse is found to be open-circuit, a new fuse can be tried, as sometimes an old fuse can deteriorate and go open-circuit. If the new one blows, there is a definite fault with the appliance and it should have professional attention; do not fit a fuse with a higher rating unless the former one was too low for the rated current of the appliance.

Check all mains leads, especially those that get frequently moved. Old leads can be a hazard, as the rubber insulation of the wires can perish, go hard, and crack off. A short-circuit could then ignite the cable outer covering, which with older cables was frequently cotton. With newer leads, watch out for heat damage. Plastic cables can melt if touched by a flat iron, curling tongs, soldering iron, or contact with a hot light bulb, and the damaged area is then liable to short-circuit. Never run mains cables under carpets. If a supply is needed on the opposite side of the room from the outlet socket, run the cable around the walls, preferably clipping it to the skirting board with cable clips. Better still, get another socket installed there.

If the above precautions are taken there should be little if any danger from electrical fires.

Open solid fuel fires for heating are less common now, modern ones are usually enclosed. Open fires in their day were responsible for many disastrous fires. A spark could pop out on to furniture or the hearth-rug, or a blazing coal could settle into an unstable position and roll off the fire on to the rug. Clothes placed to air or warm in front of the fire were a frequent cause of conflagration by slipping off the clothes-horse into the fire. If you do have open fires, all these points need to watched. A fire guard of a free-standing metal mesh should be placed in the grate around the fire whenever it is left unattended, and any remaining coals removed at night before retiring.

Another possible cause is when fire-lighting devices such as matches or a cigarette lighter are left in a place accessible to young children. All such should be kept well out of their way. An obvious point perhaps, but many tragedies have been caused by ignoring it. Usually it is a moment's thoughtlessness by putting a box of matches down somewhere handy then forgetting it when the attention is distracted.

Oil heaters were once popular as a cheap form of heating before oil prices skyrocketed. They were a great fire hazard, and also a health hazard. Perhaps it is just well they have fallen out of favour. The best advice regarding these is to scrap them!

Some hobbies or other activities may involve the use of heat or a naked flame. If these can be pursued outdoors or in an outhouse well and good, but if not, take extra precautions. If called away, it is better to turn out a bunsen burner and re-light it on return rather than let it burn unattended. Especially is this so if there are children or animals around. Dogs and cats can be very curious, and may investigate when your back is turned, to see just what it is that takes so much of your attention – attention which you could be giving to them!

The progress of a fire

Apart from instant flash fires such as those caused by chip pans, most fires go through various stages. First there is the incipient stage. This can go on for hours and starts from the time of initial ignition, perhaps just a glow or a spark starting some inflammable material smouldering, to the time when the fire gets under way. During this stage the fire can be confined and the danger is low if it is detected and dealt with. The second stage is when smoke is produced. The third stage is when flames begin to be visible and to spread. The fourth and final stage is known as the high heat stage when the fire is raging and intense heat is being produced (see Figure 66).

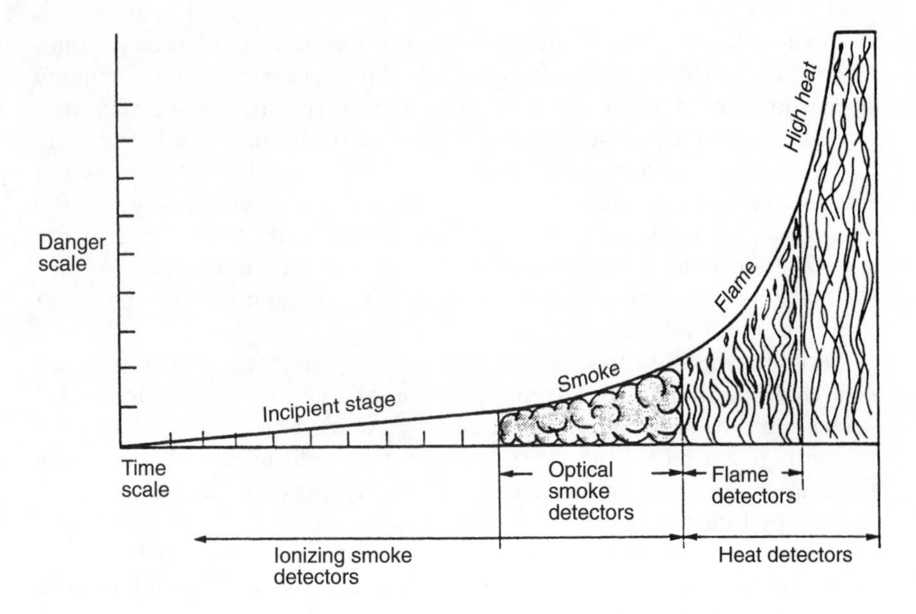

Figure 66 *The various stages of a fire and the reaction of different types of detector.*

Detectors

Fortunately elaborate fire alarm systems such as used in large buildings are not necessary for a private home. While the precautions described earlier in the chapter should prevent a fire starting, it is wise to have adequate warning if something unforeseen does happen. Time is all important and can mean the difference between life or death, so the earlier the alarm sounds the better.

Although one type of detector is most usual in domestic alarms, we will take a look at the others so as to appreciate why they are used and where, and the reasons one type is more suitable for domestic applications.

The *optical smoke detector* uses what is known as the Tyndall effect, whereby light is scattered by particles of smoke. A small radiator directs a pulsed beam of infra-red light at an angle into a detection chamber, while an infra-red detector 'looks' into the chamber. As the chamber is matt black inside, no infra-red is reflected and the detector sees nothing. If smoke enters the chamber, the infra-red is scattered in all directions by the smoke particles and some is seen by the detector. The first few pulses are ignored to prevent false alarms from random dust particles, but any after that trigger the alarm circuit.

The *ionization smoke detector* is another type. This has a small low-level radioactive radiation source such as americium 241 which maintains a flow of ions through the air in the detection chamber to a pick-up electrode. Smoke particles entering the chamber pass between the source and the electrode and reduce the flow. The drop in electrode current is sensed by a comparator circuit and triggers the alarm.

Infra-red beam detection is another type. The lamp and the detector face each other at a high level where the beam will not be interrupted by any object or person. Smoke reduces the amount of infra-red reaching the detector and an alarm is sounded when the circuits sense a drop. A delay in the response is usually incorporated to ignore momentary interruptions that could give a false alarm.

Other detectors rely on heat, one of which is the *fixed temperature detector*. This is normally set to operate at (135 °F 57 °C) at which temperature the alarm is triggered. Another is the *rate-of-rise heat detector* which triggers when the temperature climbs rapidly, typically 22 °C (40 °F) per minute. The latter usually includes a fixed temperature limit as well, so that a slow climb to a fixed temperature will also trip the alarm. These detectors can either be mechanical, using a bimetallic strip, or electronic, using a thermistor.

Another type is the *flame detector*. This responds to infra-red or ultra-violet radiation given off by the flames. The infra-red type distinguishes between flames and other sources of infra-red such as heaters or persons by sensing flame flicker. Steady sources are thus ignored. The ultra-violet type responds only to radiation in the 200–270 nm band; these frequencies in solar radiation are absorbed by the ozone layer so the sensors are not affected by sunlight.

Which detector?

Smoke is usually generated long before there is a major temperature rise (see Figure 66). As speed is vitally important in sounding the alarm, smoke detectors are preferred for most situations except those where gas, spirits or solvents may be the cause. These generate little smoke but quickly produce flame, so flame detectors are best for them.

Another situation that would eliminate the optical smoke detector is where there is a lot of dust or other solid pollutant in the air. The detector would see this as smoke and so give false alarms. Steam also can have the same effect. Furthermore, the particles would build up deposits in the detection chamber which would need frequent cleaning.

Soft furnishings and wood are the most common combustibles in the home. These generate a lot of smoke, so the smoke detector is the best for most domestic situations. The kitchen and the garage are exceptions. In the kitchen steam is often produced which could set off a smoke alarm, and fat or cooking oil, which produce instant flame, are the most frequent cause of fire. So, a flame detector is the preferred type here. In the garage also, smoke may be generated by a car starting up, and a fire more likely to be caused by petrol or oil. A flame or heat detector is therefore the best there.

A good, all-round, general-purpose detector is the ionizing smoke detector. It gives rapid response to smoke, but also will respond to the invisible products of fat or oil fires. While the right type of detector for the situation can be chosen in industrial plant, a compromise must be effected in the home where there could be several types of fire. Most of those sold for domestic use are of this type. Some have a dual chamber arrangement with comparator circuitry that reduces the possibility of false alarms due to variations in humidity.

Installation

The installation of domestic fire alarms is easy, there are no wiring or control units, as the units are self-contained and are battery operated. It is just a matter of selecting the position and then fixing to the ceiling.

Ideally a smoke alarm should be fitted in every room, but this is not often done although the cost is small for the added protection. One should though be fitted at each level. So in a two-storey house, one would be fitted in the downstairs hall, and another at the top of the stairs. If there are three storeys another should be installed in the top storey too. For split-level storeys each level should have a separate alarm.

It must be remembered that smoke will not go through a closed door, so a fire in the living room will not trigger a detector in the hall if the living room door is closed. On the other hand, a closed door will often contain a fire long enough to allow the occupants to escape. This is why a detector in each room is really the best protection. As smoke rises, the detector should be fixed to a high point in the area to be covered which usually is the ceiling. As it is not known in advance where a fire will break out, a roughly central position should be selected. An off-centre position may be chosen in a room so as to be nearer to any likely cause of fire such as the TV or the fireplace if open solid fuel or gas fires are used.

Maintenance

Some fires have been reported in which a fitted smoke alarm did not work. The reason was later found to be an exhausted battery, or in one case, there was no battery fitted at all! Most alarms have some form of built-in low-battery warning. This is often an intermittent beep which will continue for several days, in some cases for up to a month. So even if the householder is away for a period, the warning will be heard on his return. When buying a smoke alarm, make sure that an adequate low-battery warning feature is included as otherwise the battery could be forgotten.

When replacing, use an alkaline battery; although more expensive than the standard one they have a long shelf life and so will last a lot longer than the standard ones.

A press-button test facility is also included in many alarms. This will sound the alarm when pressed, although not always immediately, and the sound will stop when it is released. A weekly test would be wise, but certainly no longer than monthly. The test button checks the electronics and the sounder but not the actual smoke detector. This also should be tested periodically, say every three months. The recommended way is to ignite a piece of string and hold it smouldering under the detector and a little to one side. The alarm should sound, and stop when the unit is clear of smoke.

Periodically the unit should be cleaned. Any obvious deposit of dust can be wiped off with a damp cloth, and the interior of the unit can be carefully vacuum-cleaned without allowing the cleaner nozzle to touch the detector itself. Follow the instructions supplied by the maker.

Plan ahead

A plan should be drawn up in advance as to what to do if a fire does break out. This should stipulate escape routes and exits. These would depend on the location of the fire and so more than one plan may be needed. Assume that the fire starts at night when all are in bed. During the day not everyone will be at home and someone likely will be aware of the fire and be dealing with it.

The first priority is to get everyone to an exit point. In large families, father and mother would be assigned to rouse and collect specified children. Forget possessions, they can be replaced, lives cannot. The main exit point is likely to be the front door, where a quick check should be made that everyone is present. If the fire is not threatening, it may not be necessary to actually make an exit yet, there is little point in dragging half-clothed youngsters out on a cold, wet night unless it is necessary. But ensure the door is unlocked and a rapid exit can be made. Doors should NEVER be doubled locked at night, only during the day when the house is unoccupied. Fumbling with a key may lose vital seconds if there is a conflagration. Do not open the door though unless actually leaving, because an incoming draught will intensify the fire.

If the fire has cut off access to the front door, the back door may be reachable. If the smoke detectors are working properly and sited in the right places, there should be enough warning to enable all to get to the appropriate door before the route is cut off.

Only when everyone is at the exit point should investigation to locate the fire, proceed by someone responsible, if it is not already obvious. As all the alarms are independent, the area will be indicated by which alarm was triggered. Great caution should be exercised, if it is in a room with a closed door, feel the door first, if it is warm, do not enter, the fire has got a hold and you may be met by a wall of flame. Keeping the door closed may contain the fire and prevent it from spreading until the fire brigade can arrive.

If there is no such evidence of heat, the investigator can proceed cautiously armed with a fire extinguisher. It may then be possible to extinguish the fire if it has not developed too far. Ensure that it is completely extinguished. Any fire involving the ceiling such as one starting in a light fitting, can proceed unseen along rafters above the ceiling, and could break out again later. If in any doubt whatsoever, call the fire brigade anyway. They are the experts, they may cause some damage in making sure it has not spread, but it is better to be sure than sorry.

If the worst happens, the fire has got a firm hold, and the exit may be cut off. This could happen if it has spread to the staircase, which being wood, can quickly become an inferno. In this case the only escape may be from a first floor window. This possibility should be considered in advance and it could be facilitated in two ways. Children have been seriously injured when trying to escape from an upper window, but rope ladders can be obtained which take little room and can be stored in an odd cupboard. Such could prove to be the salvation of your family. The other aid is ensuring that at least one of the upstairs windows will fully open. The so-called 'tilt and turn' type are ideal as they open completely inwards like a door, and will make an exit much easier, should it become necessary.

However, until a means of descent is available, which will be the fireman's ladder in the absence of a rope ladder, keep all doors and windows shut. This reduces the air supply to the fire and keeps it out of the bedroom, at least for a while.

The escape route may not be aflame, but dense with smoke. Smoke can and does kill just as much as the flames. It can be poisonous and can suffocate. If one is caught in such a situation, crawl to the exit, because clear air is likely to be found at floor level. Stress this point to the whole family when discussing escape plans in advance.

Finally, have some practice runs. Not of course in the middle of the night, but make them realistic by having everyone lying on their bed to start. The children will love it!

Fire extinguishers

Not all fire extinguishers are suitable for all fires. The following are the best for particular types of fire.

Type of fire	*Type of extinguisher*
Carbonaceous, paper, wood, straw, textiles, furnishings	Water, CO_2, soda acid, foam, general purpose powder
Inflammable liquid: oil, petrol, paint, tar, paint, spirits	Foam, powder, BCF, CO_2
Gas: calor, propane, natural gas	Powder, BCF
Metal: sodium, calcium, phosphorus uranium, plutonium	Metal fire powders
Electrical fires, computers	CO_2, BCF

CO_2 = carbon dioxide gas; BCF = bromochlorodifluoromethane, a vaporizing non-toxic liquid giving a clean vapour and no deposits, and so is harmless to electronic equipment and machinery.

Fire blankets that smother a fire are most effective in some cases such as kitchen conflagrations. It will be seen that CO_2 is suitable for a wide range of fires and also being a gas, it makes less mess than some of the others such as water, foam, or powder.

Every house should have a fire extinguisher and everyone should familiarize themselves with how it works. Having to read instructions when there is a fire, would waste valuable seconds and may not even be possible in smoke and darkness. But of course it is not practicable to actually try it out! Over a period of years the extinguisher will deteriorate and may need to be re-charged by the makers or discarded and replaced. Go by the maker's recommendations printed on it. If it is to be discarded it could be used for practice on a garden bonfire, but get a new one first. Store the extinguisher where everyone knows where it is, and is easily accessible, but not in a place where it could be cut off by a fire, an upstairs landing would be ideal. A fire blanket for the kitchen is also advised.

It will probably be that all these precautions will eventually prove to have been unnecessary because most people never experience a major fire in their lifetime. If so, good! Likely you will not experience a major fire *because* of your precautions. Like insurance, you pay hoping you never will have to make a claim. But if the worst happens – and it happens every day to somebody – your efforts could mean the difference between life and death.

Index